Wm Ammentorp

THE PRODUCTIVE SCHOOL

J. ALAN THOMAS

University of Chicago

John Wiley & Sons, Inc.

New York / London / Sydney / Toronto

The Productive School

A Systems Analysis Approach to Educational Administration

Copyright © 1971 by John Wiley & Sons, Inc.

All rights reserved. Published simultaneously in Canada.

No part of this book may be reproduced by any means, nor transmitted, nor translated into a machine language without the written permission of the publisher.

Library of Congress Catalogue Card Number: 70-137110

ISBN 0-471-85900-1

Printed in the United States of America

10 9 8 7 6 5 4 3 2 1

To My Mother and Father

Preface

The ideas that are presented in this book come from many sources. While I was at Stanford University, I fell under the influence of Herbert A. Simon's *Administrative Behavior*. My thinking was expanded and refined through my reading in microeconomics, as well as in educational administration. Finally, I was affected by the extensive literature in the economics of education.

The book would not have been possible without the advice and criticism of Professor Mary Jean Bowman, University of Chicago. Professor Bowman's many helpful suggestions and incisive comments have influenced earlier versions of this book, as well as papers and articles in which I tried out these ideas on various audiences. She provided a final essential contribution by her careful reading of an advanced version of the manuscript. Many of the concepts presented in the following pages are those of Professor Bowman.

I also thank my colleagues in the field of educational finance. Dean H. Thomas James of the School of Education, Stanford University, introduced me to the field and taught me to examine assumptions carefully, to define my terms, and to write concisely. Professor R. Lyell Johns of the University of Florida also provided a careful reading of the manuscript and gave some very useful suggestions and some welcome encouragement.

This material was used as a text in several of my classes at the University of Chicago. The students who were subjected to these ideas provided excellent feedback, which was sometimes encouraging, and, when the occasion warranted, critical. My debt to these students is vast; I probably profited more than they did. They came from a variety of backgrounds. The social scientists in the group criticized the economic emphasis where it seemed to be too narrow, and the practitioners forced me to address myself to the real world of educational administration. The usual disclaimer of course applies. The errors and inadequacies of this book are attributable to the author.

Finally, I thank my wife Freda and my children, Bill, Bob, Linda, Ricky, and Eric for their understanding and support. Bill provided valuable editorial assistance in the later stages of the manuscript.

J. Alan Thomas

Contents

Chapter One
INTRODUCTION

Some Reservations — 6

Chapter Two
THE PRODUCTION OF EDUCATION

Production Functions	11
A. The Administrator's Production Function (PF1)	12
B. The Psychologist's Production Function (PF2)	13
School and Environmental Effects on Learning	17
C. The Economist's Production Function (PF3)	22
Present Value Analysis	22
Rate of Return Analysis	24
Summary	30

Chapter Three
ANALYSIS OF COSTS

1. Direct and Indirect Costs	33
The Categories of Educational Costs	33

2. Social and Private Costs	34
3. Monetary and Nonmonetary Costs	35
Aggregate Cost Analysis	35
Ingredients of Educational Costs	36
The Michigan Study	37
Indirect Costs	38
Microanalysis of Cost	39
Purpose of Cost Analysis	39
The Organizational Context of Cost Analysis	40
Elements of Unit Cost	42
Enrollment and the Cost of Educational Services	45
Variations in Output Mix	50
Indivisibilities	52
Specialization	53
Variety	54

Chapter Four

THEORETICAL APPROACHES TO RESOURCE ALLOCATION

The Production Function	59
Marginal Analysis	63

Chapter Five

APPROACHES TO DECISION MAKING

Input-Output, Cost-Effectiveness, and Cost-Benefit Models	80
1. Input-Output Models	80
2. Cost-Effectiveness Models	82
3. Cost-Benefit Models	87
Linear Programming	89
Economic Aspects of Curriculum Decisions	91
Private Decisions to Invest in Education	93

Probability Models 95
Feedback Models 98
Information Systems 103
    Categories of Relevant Data 105
        1. State Agencies 105
        2. Local Agencies 106
    Intersystem Cooperation 106

Chapter Six
ALLOCATION AND BUDGETING

Applications 109
Allocation and Input-Output Analysis 111
    Typical Allocatory Procedures 111
    Cost-Benefit Analysis and Allocation 113
The Allocation of Time 115
    Implications 117
The Budget as an Allocatory Tool 118
The Budget Document 121
    1. Line Item Budgeting 123
    2. Budgeting by Organizational Unit 123
    3. Budgeting by Functional Categories 124
    4. Budgeting by Program or Performance 125

AUTHOR INDEX 129

SUBJECT INDEX 131

Chapter One

◆◆◆◆◆

Introduction

This next decade will present new and difficult challenges for those who administer our educational systems. Administrators are being required to make their institutions responsive to their publics, during a period of rapid social change. The lessons of the past several decades do not provide complete answers to these new challenges, since many people are questioning some of the assumptions under which schools have operated. They appear to be calling for an increased emphasis on the qualitative aspects of life, and to be showing dissatisfaction with material well-being as a major goal. As part of the emphasis on quality, there is a demand that society's major organizations show greater concern for people. The depersonalization of colleges and schools would be reversed, and these institutions would be aware of the needs of their clientele as well as their employees. There are also demands that the benefits of education be more equitably distributed throughout the population. These demands come at a time when education is increasingly a prerequisite for social mobility and economic self-sufficiency for individuals, regardless of their native endowment and the status of their parents.

As one prerequisite for effective leadership, administrators of schools and colleges must have a well-developed philosophy of life, and carefully developed educational purposes. However, this is not enough to ensure needed changes in educational systems. These leaders must also possess considerable knowledge about the societal context of education. Finally, they need to be experts in organizational analysis and organizational change, since good intentions are not sufficient to bring about the goals to which they aspire.

The failure of educational institutions to fully meet the expectations which society has placed on them may, therefore, reflect a crisis in means as well as in ends. Although low achievement rates in many urban and rural schools, a mechanical approach to teaching and to organization, and a tendency to dehumanize the educational process may stem in part from a leadership which has been trained rather than educated, and which has internalized inadequate goals, even the most humanistic and enlightened leader may be overwhelmed by the practical problem of moving his organization from where it is to where he would like it to be.

Educational reformers must, therefore, understand large organizations and the manner in which they operate. This is not to advocate a narrow means-oriented approach to leadership, but to stress the importance of understanding the historical, sociological, and economic aspects of our schools and colleges. For example, by revealing the importance of the informal organization and the difference between manifest and latent functions, sociology has provided insights which are essential for the success of change-oriented leaders.

The social science approach to the understanding of educational organizations has another important purpose. School and college administrators are constantly bombarded by suggestions advocating the implementation of the latest fad, which is guaranteed to solve all the problems of education. Team teaching, nongraded classrooms, and program budgets are examples of procedures which are "packaged" and sold for easy implementation. It is hardly surprising that these processes, when introduced without an awareness of their probable anticipated and unanticipated effects on the organization, are not always completely successful. Analysis of the costs and benefits which are expected from these innovations is an essential element in assuring their success. Some approaches to this type of analysis are described in the following pages.

This book presents some conceptual bases for improved decision making. It is based on the notion that the school is a complex social system, which is in a continual process of resource interchange with its environment. The central concept is productivity, or the relationship between the outcomes of education and the human and material resources which education consumes. The term "outcome" is applicable to a variety of objectives. Hence, the procedures and concepts developed in the following pages are not specific to a given set of goals or objectives.

Education will be regarded as the result of a process of production. Once we are willing to accept the notion that education is produced, we can talk about alternative methods of production. This is a great step forward; educational systems are characterized by a sameness of organization and an apparent belief that there is a single "best" method of combining resources for producing changes in the behavior of students. The acceptance of the notion that there are alternative productive methods leads to an acceptance of resource allocation as a central responsibility of administrators.

This book is addressed to educational administrators and those responsible for their preparation. Much of the underlying rationale is drawn from economics. The joining of the work of economists and educational administrators is based on a similarity in approach; economists and administrators are both concerned with decision making, the former from a theoretical and the latter from a practical point of view. The major difference is the emphasis economists place on income as an outcome of education. Educators rightfully insist on considering a much wider range of objectives.

Because this book owes a great debt to economists past and present, I turn to a brief discussion of economic thought as applied to education. Those who are interested in exploring this subject further will find the reading of the following books and articles to be enjoyable and profitable.

Economists have for many years been interested in the contribution of education to the productive capabilities of individuals, and have considered education as one factor that helps to bring about economic growth.[1] In a fine historical treatment of the subject, Mary Jean Bowman[2] pointed out that the mercantilists appreciated the idea of investment in man, but "did not follow through with their observations concerning human skills and productivity to analyze costs and returns."[3] The classical economists were also concerned with education, and in Adam Smith's *Wealth of Nations* the analogy between men and machines is, in Bowman's words, "an unambiguous anticipation of recent work."[4] While Marshall discarded the notion of human capital, Irving Fisher "brought the human component of capital fully into the fold."[5]

In recent years, economists have revised and expanded their interest

---

[1] Mark Blaug, *Economics of Education: A Selected Annotated Bibliography* (London: Pergamon Press, 1966).

[2] Mary Jean Bowman, "The Human Investment Revolution in Economic Thought," *Sociology of Education*, **39**, No. 2 (Spring, 1966), pp. 111–137.

[3] *Ibid.*, p. 113.

[4] *Ibid.*, p. 113.

[5] *Ibid.*, p. 114.

in human investment. Much of the recent work in this field has been stimulated by Theodore W. Schultz.[6] Some of the recent empirical studies incorporate sophisticated calculations of rates of return to investment in education[7] and of education's contribution to the gross national product.[8]

Most of the work done to date uses census data to relate the additional income associated with a year of schooling to the cost of this schooling. Such aggregate studies do not adequately measure the value of the various levels of education; for example, they do not satisfactorily deal with decisions concerning the allocation of resources among preschool, secondary school, and higher education.

Nor do they provide information about the profitability of the various curricula or of alternative instructional or organizational methods. Although there are a number of studies dealing with these kinds of variables, the complexity of the economy and the national educational system is such as to render any study of the total relationship between educational institutions and the national economy almost impossible. Linear programming provides a means for dealing with these problems in societies which are somewhat smaller and less complex.[9]

Microeconomics, or the application of economics to business organizations, also has important implications for educational administration. Although the foundations of this aspect of economic analysis, and especially the principle of diminishing returns to increased investment, were provided by the classical school,[10] the formal tools used in micro-

---

[6] Theodore W. Shultz, *The Economic Value of Education* (New York: Columbia University Press, 1963).

[7] Gary S. Becker, *Human Capital* (New York: Columbia University Press, 1964).

[8] Edward Denison, *The Determinants of Economic Growth in the United States* (New York: Committee for Economic Development, 1962).

[9] Samuel Bowles, *Planning Educational Systems for Economic Growth* (Cambridge, Mass., Harvard University Press, 1969).

[10] See William J. Barber, *A History of Economic Thought* (Baltimore, Maryland: Penguin Books), pp. 80–81. In his chapter on David Ricardo, Barber wrote:
> Acreages of high fertility (which, it could be reasonably assumed, would be the first to be drawn into cultivation) would provide a windfall to their owners. Moreover, the size of this windfall would swell as population growth enlarged the demand for food. As food prices rose, less fertile areas would be brought under the plough, so long as their cultivators could obtain going rates of return for their efforts. Meanwhile, the owners of fertile acreages would reap higher and higher rents. The outputs of the last units, on the other hand, would be sufficient only to cover the costs of cultivation and would fail to yield a rate.

economics were developed by the neoclassicists.[11] Recent economists, such as Samuelson[12] and Baumol,[13] have developed mathematical models of the internal economies of organizations.

Some preliminary attempts have been made to conduct empirical studies which can be used for improving decision making in educational organizations. Kershaw and McKean pioneered the field, by proposing the use of multivariate analysis to determine the marginal effects of various inputs on the efficiency of educational organizations.[14]

Burkhead conducted an empirical study in four cities to attempt to discover the important input-output relationships.[15] In an important but little known California study, Benson *et al.* studied the relationship between selected inputs and outputs, and proposed a decision procedure using feedback from achievement scores as a partial basis for the allocation of resources to low-achievement local school districts.[16]

Teachers' salaries constitute a major part of most school budgets. An examination of the methods of remunerating teachers is, therefore, central to any cost-benefit analysis of educational decisions. Kershaw and McKean pointed out that teacher shortages differ according to subject field, and that a method of remuneration based on the market demand for different categories of teachers would be more helpful in eliminating these shortages than the single salary schedule, which pays teachers according to their experience and qualifications.[17]

In the absence of knowledge about costs and benefits associated with salary schedules, their structure is determined largely by political pres-

---

[11] Alfred Marshall, *Principles of Economics* (London: Macmillan, 1925).

[12] Paul A. Samuelson, *Foundations of Economic Analysis* (Cambridge: Harvard University Press, 1947).

[13] William J. Baumol, *Economic Theory and Operations Analysis* (Englewood Cliffs, New Jersey: Prentice Hall, 1961).

[14] J. A. Kershaw and R. N. McKean, "Systems Analysis and Education," Research Memorandum RM-2472-FF (Santa Monica, Calif.: The RAND Corporation, October 30, 1969).

[15] Jesse Burkhead, *Input and Output in Large City High Schools* (Syracuse: Syracuse University Press, 1967).

[16] Report of the Senate Fact Finding Committee on Revenue and Taxation, *State and Local Fiscal Relationships in Public Education in California* (Sacramento: Senate of the State of California, 1965). Report prepared by Charles S. Benson *et al.*

[17] Joseph A. Kershaw and Roland N. McKean, *Teacher Shortages and Salary Schedules* (New York: McGraw-Hill, 1962).

sures. Harvey[18] and Wilkinson[19] have both studied salary schedules in Canada, using cost-benefit analysis. Their findings have important implications for policy. Pedersen studied salary schedules as a variable which has an effect on decisions by teachers to move from one school system to another.[20]

Some of the most creative work in the economics of education is being carried out at the University of Chicago by Professor Mary Jean Bowman. As theorist,[21] critic,[22] methodologist,[23] and economic historian,[24] she is able to bring together a knowledge of the institutional, sociological, and economic aspects of education.

SOME RESERVATIONS

There is some similarity between the ideas presented in this book and those of the scientific management school which flourished in the early 1900s. This era was chronicled by Raymond E. Callahan in *Education and the Cult of Efficiency*.[25] Callahan told how the cost-conscious critics of the schools were successful in causing school administrators to implement a variety of procedures, including cost analysis and the widespread use of achievement testing. Efficiency experts with their stopwatches came into the schools to evaluate teachers as well as janitors.

---

[18] Valerien Harvey, "Economic Aspects of Teachers' Salaries," Ph.D. dissertation, Department of Education, University of Chicago, 1967.

[19] Bruce William Wilkinson, *Studies in the Economics of Education*, Department of Labor, Canada, Occasional Paper No. 4 (Ottawa: Queen's Printer, 1966).

[20] K. George Pedersen, "Selected Correlates of Teacher Migration," Ph.D. dissertation, Department of Education, University of Chicago.

[21] Mary Jean Bowman, "Human Capital: Concepts and Measures," in *The Economics of Higher Education,* edited by Selma Mushkin (Washington: U. S. Department of Health, Education and Welfare, Office of Education, 1962).

[22] ———, "Schultz, Denison, and the Contribution of 'Eds' to National Income Growth," *Journal of Political Economy,* **LXXII**, No. 5 (October, 1964), pp. 450–464.

[23] ———, "The Costing of Human Resource Development," in *The Economics of Education,* edited by E. A. G. Robinson and John Vaizey (London: Macmillan and Company, 1966).

[24] ———, "The Land Grant Colleges and Universities in Human Resource Development," *The Journal of Economic History* (December, 1962), pp. 523–546.

[25] Raymond E. Callahan, *Education and the Cult of Efficiency* (Chicago: University of Chicago Press, 1962). Copyright 1962 by The University of Chicago.

There was also a great emphasis on such aspects of administration as the choice of the best floor wax, and the determination of the optimal amount of floor space for which janitors should be responsible. However, Callahan points out:

> the essence of the tragedy was in adopting values and practices indiscriminately and applying them with little or no consideration of educational values or purposes. It was not that some of the ideas from the business world might not have been used to advantage in educational administration, but that the wholesale adoption of the basic values as well as the techniques of the business/industrial world, was a serious mistake in an institution whose primary purpose was the education of children. Perhaps the tragedy was not inherent in the borrowing from business and industry but only in the application. It is possible that if educators had sought "the finest product at the lowest cost"—a dictum which is sometimes claimed to be a basic premise in American manufacturing—the results would not have been unfortunate. But the record shows that the emphasis was not at all upon "producing the finest product" but on "the lowest costs."[26]

The lesson which Callahan reveals in *Education and the Cult of Efficiency* must not be forgotten. In particular, procedures developed in business and industry cannot be translated, without change, to activities in the public sector. This book attempts to resolve the problem by developing a mode of analysis that is specific to education, and that emphasizes outcomes as well as costs. The analysis is generalizable to decisions based on nonquantitative as well as quantitative evidence. Furthermore, it would be difficult to refute the following statements:

1. Individuals must make choices—whether in their private lives or in their capacity as members of organizations.
2. So far as possible, choices should be conscious.
3. Choices should be based on all available information, both quantified and nonquantified.
4. This information should include data about both the potential benefits to be gained by the selection of an alternative and the costs of that alternative. Both benefits and costs should include nonquantified psychological factors, as well as so-called "hard data." After he has examined this evidence, the individual who wishes to increase his total satisfaction will make a decision based on the costs and the benefits associated with each alternative he considers.

---

[26] *Ibid.*, p. 244.

My first task will be to build a conceptual base from which implications for practice may be drawn. With this goal in view, Chapter 2 discusses the implications that may be drawn from considering schools as open systems, and examines ways in which the inputs and outputs of educational systems may be studied. Costs constitute a special kind of input of particular importance to this book; the analysis of costs is discussed in Chapter 3. These concepts are drawn together in a set of analytic procedures in Chapter 4. The last two chapters deal with implications for practice. Chapter 5 treats decision making, and Chapter 6 treats a special kind of decision making, namely, choosing from among alternative forms of resource allocation.

Note regarding use of mathematics. The models in this book can often be expressed more clearly through the use of elementary algebra and geometry. Where this is the case, I have not attempted to dispense with the use of elementary mathematics. The use of such tools is, I believe, essential in the development of improved managerial techniques in education, as in other fields. Those readers who wish to study these problems in greater depth may wish to establish skills in linear algebra, basic economics, and operations analysis. Baumol (*op. cit.*) provides an excellent supplementary text.

Chapter Two

The Production of Education

The central responsibility of the administrator is to create and operate a productive system. This implies that he must use the resources at his disposal to achieve, as fully as possible, the goals of the system. It also suggests that he must monitor the system, using information about its performance at a given point in time in order to improve its subsequent operation.

This term "system" is much in vogue, in education and in many other fields. Its primary connotation is interrelatedness; a system is a set of parts that are related to each other.[1] Man-made systems are devised to achieve a given purpose or set of purposes. Educational systems consist of interrelated components (people, buildings, books, and equipment) and are constructed to bring about intended changes in the behavior of the client.

Systems designed to bring about changes in the behavior of large numbers of people are by necessity very complex. Also, the results of given procedures carried out within educational systems can seldom be

---

[1] There is an extensive literature on systems. Our recommendation for a companion reference to this book is Stafford Beer's *Cybernetics and Management* (New York: Wiley, Science editions, 1964). Also, we recommend Beer's more advanced text, *Decision and Control* (Wiley, 1964).

accurately predicted. Uncertainty is magnified by the fact that the learner is also affected by his out-of-school environment. Educational systems and the surrounding environment are constantly interacting. Hence, such systems are best described as *open,* in contrast to closed, machine-like systems.

Open systems receive support from their environment, and return products of greater or less usefulness to their surroundings. Thus, the environment provides the system with *inputs* and receives its *outputs*. When these inputs and outputs can be identified and measured, they provide useful information about the system. Some systems are so complex or so inaccessible that they can only be studied through input-output analysis.[2]

Input-output analysis is particularly useful in the study of the complex social systems we call organizations. Such study is complicated by the fact that organizations serve many purposes; and the output must be redefined as a new purpose is considered. For example, General Motors may be thought of as a system designed to produce automobiles. However, its shareholders may consider its main output to be a financial profit. To its workers, its output consists of their salary checks. Young single office workers may think of General Motors as a social organization, or even a matrimonial bureau. Similarly, the outputs of schools may consist of additions to students' ability to enjoy and appreciate life, or, from an altogether different perspective, of increments to the national income.

Whatever output we select for study may be considered in relationship to the corresponding inputs. Since outputs are valued and inputs are scarce, the ratio of outputs to inputs should be maximized; a *productive organization* is one with a favorable balance of outputs to inputs. The mathematical relationship between outputs and inputs can be expressed as an equation called a *production function.*

The concept of production function should, in my opinion, be added to the literature of educational administration. The concept serves several purposes. In the first place, the concept emphasizes that there is more than one way to produce education. From kindergarten through graduate school, educators are overinclined to follow certain instructional and organizational procedures simply because they are sanctioned by common practice. Ratios of students to teachers and to classrooms, books, and counsellors are often formalized and applied with little or no empirical or theoretical justification. The production function concept treats

---

[2] For example, geologists use input-output analysis with seismographic studies to study the earth's crust, while medical scientists sometimes use similar methods to study dysfunctions in the human heart.

these ratios as variables rather than constants, and forces attention on alternative methods of allocating scarce resources.

In the second place, the production function provides a mathematical basis for decision-making models. These decision models are intended to provide a means for choosing from among alternative solutions to a given problem. In effect, a new dimension of administrative theory, based on economics, is introduced by those who follow these procedures. This theory becomes useful also when applied to an evaluation of an entire educational system, or of curricula, teaching methods, or organizational procedures within the system. Finally, these procedures provide a conceptual basis for budgetary practice, which is particularly useful when program budgeting is introduced.

PRODUCTION FUNCTIONS

It is easy to become confused by the various concepts used to describe the resource interchange between educational systems and their environment. Such terms as cost-benefit and input-output analysis are often used synonymously, and both are confused with cost-effectiveness analysis. There is sometimes a vague belief that these concepts are related to program budgeting (or PPBS, to use the current jargon), but the relationship is often not specified. This confusion results from the fact that the production of education can be viewed in a number of ways. In this chapter the *production function* concept is used as a basis for examining three distinct types of input-output relationships.[3] They are

---

[3] The word "function" is used in its mathematical and not its sociological sense. It is defined as a relationship between one or more inputs and a selected output. The form of the statement is therefore:

$$O_i = g\,(I_1, I_2, \ldots I_n)$$

where O_i is the "ith" output, and the I's are the inputs.

It is sometimes necessary to consider outputs as being jointly produced. That is, a given set of inputs produces more than one output simultaneously. (For example, teaching and research, or reading skills and good citizenship are produced jointly.)

See Andre Daniere, *Higher Education in the American Economy* (New York: Random House, 1964). In this case,

$$(O_1, O_2, \ldots O_n) = g\,(I_1, I_2, \ldots I_m)$$

This expression is, of course, more difficult to deal with than the single output case.

distinguished from each other through the manner in which inputs and outputs are defined. The production functions are named for their principal users, the administrator, the psychologist, and the economist.

A. THE ADMINISTRATOR'S PRODUCTION FUNCTION (PF1)

The school or college administrator has the responsibility of developing an educational system. His typical procedure is to identify a set of courses and other services that are in demand by his students or their parents (as in the case of colleges), or that are defined by teachers, administrators, and legislators as "needed" by students. In order that these services may be provided, he must obtain inputs—units of space, people, books, and equipment. He then combines these inputs with groups of students and with the kinds of curricula that have been defined as appropriate.

Given the complexity of large educational agencies, the tendency to regard services (including academic courses) as outputs is an understandable one. Furthermore, the quality dimension is not ignored. Administrators may interpret the quality of an educational system as a function of the quantity and quality of inputs, including, for example, class size, teachers' qualifications, building construction, the size and contents of the library, and equipment in the science laboratories. Since input quality is closely related to program cost, cost and program quality are sometimes assumed to be identical.

In this production function, then, outputs are defined as units of specific services. The output unit must include a time dimension, such as a student-year, or a student-hour, in order that the costs of providing different outputs may be compared. Inputs include goods that are purchased and people who are hired to provide these services. Among the significant inputs in educational systems are space, equipment, books, materials, and the time of teachers and other employees. These inputs are purchased with money—it is, therefore, possible to calculate the cost per student-hour of providing a given service.

An analysis based on PF1 does not require elaborate mathematical or statistical calculations. The computation of the cost of providing a unit of a given service is an accounting procedure, and may be routinized and programmed for the computer. These kinds of analyses have important practical and research implications. While there are few cases where decisions should be made *entirely* on the basis of the cost of providing a unit of a given service, cost data are an essential administrative tool, to be used in conjunction with other data, including feedback about outputs.

Accurate cost data are essential for perfo᠎ ᠎ing research into administrative problems in education. If, for example, a school system is considering whether to construct two high schools or only one, the cost implications of the decision should be taken into account. It is rather remarkable, to take another example, that so little is known about the monetary cost associated with maintaining so many small school districts throughout the nation. Finally, statistical input-output studies cannot lead to recommendations for action unless the cost implications of substituting one input for another are known.

B. THE PSYCHOLOGIST'S PRODUCTION FUNCTION (PF2)

I have given the above title, somewhat arbitrarily, to the function whose outputs are behavioral changes in students, including additions to knowledge, the acquisition of values, or the increased ability to relate to others. These outcomes are in the domain of the psychologist who may, however, be less than enthusiastic about the kinds of input variables used in this analysis.

There is an implied relationship between the first and second production functions, since the services provided in schools are designed to produce changes in behavior. However, decisions regarding the first production function are made on the basis of the cost of producing a given set of services. Decisions concerning PF2 may be made on the basis of statistical studies which treat both inputs and outputs as variables. The basic methodology in PF1 is cost accounting; that in PF2 is multivariate statistical analysis.

A major difficulty in this analysis is identifying and measuring *changes* in behavior. Ideally, inputs should be related to "value added" or to increments in achievement, however achievement is defined. Thus far, most studies have used measures of behavior as the output variable, with the implicit assumption that these measures can be taken to represent achievement increments. Furthermore, most studies to date have used cognitive outcomes as the criterion. These outcomes are relatively easy to measure. However, outputs in the affective domain should also be considered; such variables are related to the long-term results of education, including those which the economist is interested in.[4]

Inputs in PF2 include the time of teachers and other employees, as well as space, books, equipment, and materials. Qualitative attributes of teachers, including their intellectual capabilities and their attitudes toward students are also relevant. The fact that studies to date have concentrated on the more overt characteristics of schools and have avoided these more subtle psychological variables is a limitation which can be remedied in future studies.

An important input in PF2 is the time of students, since it may be assumed that the amount of time students spend studying a given subject will help determine their performance. Equally important, characteristics of students, such as their aptitudes, interests, and motivations affect output levels. Where these variables cannot be measured directly, related variables, including characteristics of the home and community, can be used as proxies.[5]

The total effect of all the inputs on outputs, as well as the separate effect of each input (with other inputs held constant), can be studied by means of multivariate analysis. A common procedure has been to gather data from a large number of educational systems, and to subject this data to multiple regression analysis. The regression coefficients which are assigned to each input variable may be used to help guide resource allocation. However, the "association type" study is a poor indicator of cause and effect. It would be much better if longitudinal studies could be carried out which would document changes in outputs associated with changes in given inputs. Such studies are extremely difficult to perform, because of the time involved, and because other changes which take place over time may cloud the results.

---

[4] One study which included affective as well as cognitive outcome is Mark Holmes' "Prediction of Educational Outcomes from Institutional Variables in a Population of Junior High Schools," Ph.D. dissertation, University of Chicago, 1969.

[5] See, for example, the studies performed by students of Benjamin S. Bloom, and reported by Robin H. Farquhar in "Home Influences on Achievement and Intelligence: An Essay Review," *Administrator's Notebook*, **XIII**, No. 5 (January, 1965); Benjamin S. Bloom, *Stability and Change in Human Characteristics* (New York: Wiley, 1964). The two most relevant studies are Ravindrakumar H. Dave, "The Identification and Measurement of Environmental Process Variables That Are Related to Educational Achievement," Ph.D. dissertation, Department of Education, The University of Chicago, 1963; and Richard M. Wolf, "The Identification and Measurement of Environmental Process Variables Related to Intelligence." Ph.D. dissertation, Department of Education, the University of Chicago, 1964.

One interesting area of application of PF2 has been the investigation of the effect of school or school system size on learning. The assumption behind such studies is that large systems might (through specialization and the division of labor) have a greater effect on students' behavior per dollar spent than smaller systems. These studies are not definitive, due largely to their failure to make adequate use of cost data. Nevertheless, they point to a second way of using scale of operation as a variable in educational research.[6]

Some exploratory work has been done toward using input-output analysis for evaluating educational systems.[7] If the appropriate background variables can be controlled, it may be possible to identify the low- and high-productivity schools.

The procedure is as follows. First, a fairly large number of schools or school systems is chosen by some process of random sampling from a given population. Tests of achievement are administered in each school or school system. Also, census-type data are gathered from the environment of the educational system. These data should include information about income, parental education, and occupational status. Through regression analysis, the relationship between achievement and the background variables is analyzed. As a result of this analysis, a regression line, relating achievement to a set of background variables, is drawn (Figure 2-1).

Each school or school system may now be placed on the graph. Some will fall on the line; others will be above or below it. Some relatively low status schools will achieve at a high level; this success may be attributed in part to the school. Some high status schools will fall below the regression line; it may be hypothesized that these schools may be partially at fault. Strategies for using this analysis for improving performance, such as relating each school in a city school system to the regression line and setting merit-related salaries of principals accordingly, may be devised.

There are, however, flaws in this procedure. One problem that will be encountered is disagreement with respect to the outputs which should be considered. At the least, evaluation should be based on a comparison of schools which have accepted similar objectives. Another problem is that the equation which depicts the input-output relationship will contain

---

[6] Elchanan Cohn, "Economies of Scale in Iowa High School Operation," *Journal of Human Resources,* III, No. 4 (Fall, 1968), pp. 422–434.

[7] See, for example, New York State's interesting work along this line. Samuel M. Goodman, *The Assessment of School Quality* (Albany, N. Y.: The University of the State of New York, 1959).

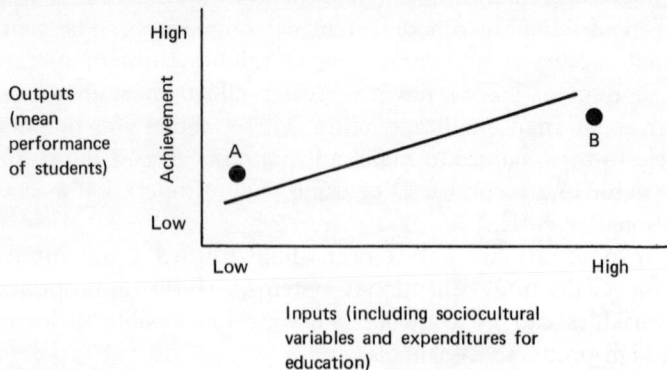

Figure 2-1 Regression of background variables in achievement.
Note: Even though the performance in School B is superior to that in School A, the former is under productive, and the latter is highly productive, since A is above the regression line, while B is below it.

a larger error variance. Conclusions based on statistically insignificant variations from the regression line should be avoided.

Among the first to advocate the use of input-output analysis in education were two economists from the RAND Corporation, J. A. Kershaw and R. N. McKean (*op. cit.*). They proposed gathering input and output data from a large number of schools, and subjecting the results to covariate analysis. They proposed achievement test scores as output measures, and information about such school characteristics as the size of the library, the availability of counsellors, class size, teachers' qualifications, and per pupil expenditures as inputs. The resulting analysis would be used as a guide in the improvement of resource allocation within educational systems; resources would be shifted from the inputs that contribute least to achievement to the inputs that contribute most. Their proposal, while novel and imaginative, has at least two difficulties in it.

First, the use of achievement data as output measures has certain limitations. It places an emphasis on the quantifiable outcomes of education as opposed to less tangible, but equally important, objectives. Furthermore, this is not a completely satisfactory output in terms of open systems theory, because of the limited relationship between school achievement and contributions to the wider society such as income and employability. Finally, it is often claimed that achievement measures are inaccurate, and contain sampling and measurement errors. Nevertheless, these measures do predict with fairly high accuracy both future success in school and the tendency to obtain additional education. The amount of schooling that people possess is, in turn, related to future income and employability.

Second, in the absence of a general theory of instruction, indices of the effect of the school on learning tend to include a mixed bag of variables. Since the ability of these variables to predict output levels is limited, there is inadequate empirical evidence for developing a taxonomy of school effects. Two further limitations are noted. In the first place, most input-output models used to date have not included costs. Since decisions must be partially based on cost, this omission needs correcting. In the second place, the initial concepts of input ignored out-of-school effects on learning. Hence, a modification of the original design proposed by Kershaw and McKean is necessary.

School and Environmental Effects on Learning

Scholars in the field of sociology and human development have long emphasized that learning and motivation are heavily affected by out-of-school influences.[8] One major problem for systems analysis is to determine the nature of the interaction between the effects on achievement of (a) the school and (b) children's backgrounds.

To oversimplify the issue, we identify three types of models which appear to be implicit in various analyses of the problem.

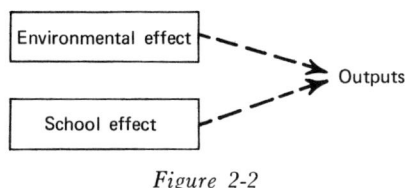

Figure 2-2

1. *An additive model* (Figure 2-2). According to this model, the environmental and school effects are separate, but additive. The challenge presented by this model is to separate the effects of the school from those of the community. In actual practice, this separation can only be carried out when either the school or the

[8] See for example Robert J. Havighurst, *Growing Up in River City* (New York: Wiley, 1962).

community (or home) has complete custody of the child. However, statisticians have devised ways whereby certain variables can be controlled or held constant, while the relationship between the noncontrolled independent variable and the dependent variable is studied. The use of this technique in input-output analysis is reported at a later point in the chapter.

2. *The open systems model* (Figure 2-3). This more realistic approach regards the environment as influencing the nature of

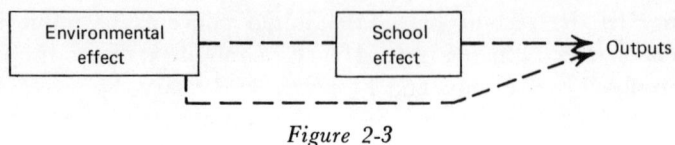

Figure 2-3

school system inputs. The school, through its various activities, processes these inputs into outputs. Research conducted on the basis of this model is not concerned with holding certain effects constant while studying others. Rather, all variables are examined simultaneously. Another type of study that is useful is a longitudinal analysis of changes in inputs, processes, and outputs in a single educational system. Finally, experiments involving different combinations of inputs provide an appropriate way of obtaining new knowledge about the effectiveness of alternative treatments with different client groups.

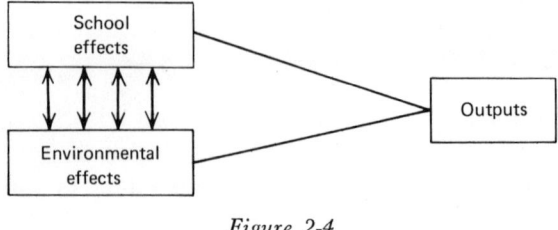

Figure 2-4

3. *The interactive model* (Figure 2-4). A third model, more adequate as an explanatory device, but extremely complex, is what might be called the "interactive model." This model recognizes the interaction between school, home, and community, and the

relationship between this interaction and the quality of outputs. The mathematical procedure implicit in this model has not yet been developed. It appears to involve the solution of a complex set of simultaneous equations.

The most ambitious single effort to determine the nature of input-output relationships was conducted by James S. Coleman for the U. S. Office of Education.[9] The main purpose of this study was to examine the nature of educational opportunity in the United States, especially with regard to racial differences in the student body.

Only a portion of the Coleman Report is related to the effect of school characteristics on achievement. However, this aspect of the study has been given a great deal of publicity, and some of the findings are now accepted as a basis for policy. Coleman reports as follows:

> When one sees that the average score on a verbal achievement test in school X is 55 and in school Y is 72, the natural question to ask is: What accounts for the difference?
>
> There are many factors that may be associated with the difference. This analysis concentrates on one cluster of those factors. It attempts to describe what relationship the school's characteristics themselves (libraries, for example, and teachers and laboratories, and so on) seem to have on the achievement of majority and minority groups (separately for each group on a nationwide basis, and also for Negro and white pupils in the North and South).
>
> The first finding is that the schools are remarkably similar in the way they relate to the achievement of their pupils when the socio-economic background of the students is taken into account. It is known that socioeconomic factors bear a strong relation to academic achievement. When these factors are statistically controlled, however, it appears that differences between schools account for only a small fraction of differences in pupil achievement.
>
> The schools do differ, however, in their relation to the various racial and ethnic groups. The average white student's achievement seems to be less affected by the strength or weakness of his school's facilities, curriculums, or teachers than is the average minority pupil's. . . .
>
> The inference might then be that improving the school of a minority pupil may increase his achievement more than would improving the school of a white child increase his.[10]

These conclusions are based in part on data reported in Table 2-1.

---

[9] James S. Coleman, *Equality of Educational Opportunity* (Washington, D. C.: U. S. Department of Health, Education, and Welfare, Office of Education, 1966).

[10] *Ibid.*, pp. 21–22.

20 THE PRODUCTION OF EDUCATION

Table 2-1 Percent of Variation Explained in Scores on Four Tests by Selected School Characteristics for Negroes and Whites, After Student Background Effects are Controlled, for Grade 12.

	Negroes				Whites			
	Verbal	Non-Verbal	Reading	Math	Verbal	Non-Verbal	Reading	Math
10 teacher variables	7.21	5.21	4.46	2.06	1.29	1.19	0.23	0.61
18 school variables	6.54	5.78	5.00	2.12	2.02	0.81	0.62	0.90

Source: James S. Coleman, *Equality of Educational Opportunity* (Washington, D.C.: U. S. Department of Health, Education and Welfare, Office of Education, 1966), p. 294.

Samuel Bowles and Henry M. Levin have seriously criticized the study and its findings.[11] Some of their criticisms were:

1. The measurement of school resources was inadequate. For example, the study used average expenditure per pupil in the district, and did not take into consideration differences in expenditures among schools, or even differences between the level of expenditures in elementary and secondary schools.
2. Only a very limited range of variables was used to determine the effect of school facilities.
3. The study collected information only on current school inputs, and neglected the cumulative effect of past expenditures on achievement.

Their most important criticism, however, centers around the procedure of holding background variables constant, and then examining the school effect on the unexplained variance in achievement. Since the school effect and background effect variables are intercorrelated, this has the effect of unnecessarily reducing the apparent effect of school variables. They suggest that a better approach is to use the regression

---

[11] Samuel S. Bowles and Henry M. Levin, "The Determinants of Scholastic Achievement, an Appraisal of Some Recent Evidence," *The Journal of Human Resources,* III, No. 1 (Winter, 1968), pp. 3–24. Also, *op. cit.* See also Marshall S. Smith, "Equality of Educational Opportunity: Comments on Bowles and Levin," and Glen G. Cain and Harold W. Watts, "The Controversy about the Coleman Report: Comment," both in *Journal of Human Resources* (Summer, 1968), pp. 384–389 and 389–392.

coefficients, and they report the following equations:

Regression Equations for Negro Verbal Achievement at Grade 12

X_1 = Negro student's verbal score
X_2 = Reading material in home
X_3 = Siblings (positive equals few)
X_4 = Parents' education
X_5 = Science lab facilities
X_6 = Teacher salary (in thousands of dollars)
X_7 = Teacher verbal score

(1) $X_1 = 33.40 + 1.99X_2 + 1.86X_3 + 2.49X_4 + 0.062X_5 + 1.78X_6$
    (2.66) (4.34) (4.49) (3.18) (5.95)
$R^2 = 0.1506$
$|X'X| = 0.68$ = Measure of collinearity

(2) $X_1 = 19.49 + 2.09X_2 + 1.81X_3 + 2.42X_4 + 0.050X_5 + 1.24X_7$
    (2.82) (4.25) (4.38) (2.58) (7.14)
$R^2 = 0.1633$
$|X'X| = 0.68$ = Measure of collinearity

Note: The *t*-values appear in parentheses. Definitions of the variables appear in Volume II of the Coleman Report, pp. iii–vii.
Source: Samuel S. Bowles and Henry M. Levin, "More on Multicollinearity and the Effectiveness of Schools," *The Journal of Human Resources,* III, No. 3, (Summer, 1968), pp. 393–400.

Far from suggesting that schools make no difference, these equations show the quality of teachers, as reflected in their salaries and their verbal scores, to be an important variable in affecting student achievement. Although no comprehensive measure of school facilities was available to the researchers, the measure they did use (presence of science labs) is also significantly related to output. (Remember that these relationships hold in the presence of the environmental variables which were selected.) Other studies have also indicated that teacher quality is a statistically significant predictor of achievement.[12,13,14]

[12] Alan Thomas, "Efficiency in Education: An Empirical Study," *Administrator's Notebook,* **XI**, No. 2 (October, 1962).
[13] Jesse Burkhead, *Input and Outputs of Large City High Schools* (Syracuse: Syracuse University Press, 1966), pp. 56, 72–81.
[14] Charles S. Benson, *Report of the Senate Fact Finding Committee on Revenue and Taxation, State and Local Fiscal Relationships on Public Education in California* (Sacramento: Senate of the State of California, 1965), p. 58. "We are led to the conclusion that the caliber of teachers is the single most important factor."

C. THE ECONOMIST'S PRODUCTION FUNCTION (PF3)

The economist sees education as contributing individuals with acquired competences to the economic system.[15] In return, the economy contributes resources for the operation of schools. A productive school is one for which the monetary value of the education individuals receive is in a favorable balance to the cost of providing that education.

In this production function, outputs are the additional earnings which result from an increment of schooling, while inputs comprise the cost of that increment. An additional cost element, not present in the preceding production functions, is the opportunity cost of students' time.

The major advantage of the third production function lies in the refined analytic tools developed by economists. These tools were devised for measuring the productivity of physical capital. However, they are equally useful in the analysis of the costs and benefits associated with the production of human capital. Two procedures are used: (1) present value analysis and (2) the analysis of internal rate of return.

Present Value Analysis

Both the costs and the benefits of education are incurred over a period of years. The benefits which individuals (or aggregates of individuals) obtain as the result of an increment of schooling take the form of a stream of income received over their lifetime. It is therefore necessary to devise methods for comparing streams of income and streams of cost. This can best be done by reducing both streams to a base year value—this is called present value analysis.

Present value analysis consists of using compound interest and compound discount procedures to reduce a stream of costs and a stream of income to their value at a given base year. A dollar received today could be invested to yield interest; if that interest is allowed to accumulate, today's dollar will increase at a compound rate of interest. But this also implies that a guaranteed payment of one dollar in 10 years time is equivalent to a smaller amount of money today. If the rate

---

[15] Gary S. Becker, *Human Capital* (New York: National Bureau of Economic Research, 1967).

of interest is i, the present value of one dollar to be received 10 years from now is

$$\frac{1}{(1+i)^{10}}$$

Similarly, the present value as of year 0 of an income stream of $1 per year for 10 years is V_0Y, where

$$V_0Y = \frac{1}{(1+i)} + \frac{1}{(1+i)^2} + \cdots \frac{1}{(1+i)^{10}}$$

or

$$V_0Y = \sum_{t=1}^{t=10} \frac{1}{(1+i)^t}$$

The present value of an income of Y dollars per year for n years beginning with year $t = 1$ is this:

$$V_0(Y) = \sum_{t=1}^{t=n} \frac{Y_t}{(1+i)^t}$$

In this case, Y may of course vary from year to year.

Costs can be treated in the same fashion. The present value at year 0 of a stream of costs amounting to $X per year is, therefore,

$$V_0(X) = \sum_{t=1}^{t=n} \frac{X_t}{(1+i)^t}$$

The value of any particular Y_t or X_t can of course be zero.

A criterion for determining whether or not an individual should continue his schooling can now be stated. An investment in an increment of education is worthwhile if the present value of the additional benefits associated with this increment is greater than the present value of the additional costs. That is, an investment is worthwhile if:

$$V_0(Y) - V_0(X) = \sum_{t=1}^{t=n} \frac{Y_t}{(1+i)^t} - \sum_{t=1}^{t=n} \frac{X_t}{(1+i)^t} > 0$$

A major problem in this procedure lies in determining the discount rate which should be used for purposes of calculations. The appropriate rate of discount depends on the time preferences of individuals. If a person prefers to receive his income as soon as possible he will utilize a high subjective rate of discount, thus reducing the present value of an income stream. Harvey has examined this issue, and concludes:

> When the expression $V_0(Y) - V_0(X)$ is evaluated, it is necessary to select a value for the external rate of return deemed "appropriate" to discount the two streams. But what is an appropriate rate to use? Ideally, this will be a benchmark or "reservation" rate such that any value of $V_0(Y) - V_0(X)$ greater than zero signals that the investment would be more profitable than any alternative. The reservation rate of discount is the highest of the following: (1) the subjective time preference of the individual, (2) rates of return available on alternative investments, and (3) in the event that he must borrow to finance the investment, the interest rate he must pay to obtain funds for such investment. In fact, it is always informative to solve . . . for several interest rates.[16]

Rate of Return Analysis

The internal rate of return to an investment is defined as that rate which equates the present value of the investment to zero.

$$V_0(Y) - V_0(X) = \sum_{t=1}^{t=n} \frac{Y_t}{(1+i)^t} - \sum_{t=1}^{t=n} \frac{X_t}{(1+i)^t} = 0$$

The decision rule is that individuals (or social groups) should continue to invest as long as the rate of return exceeds that obtainable from other alternatives which are considered.

Research techniques of even greater generality have been developed and applied to the analysis of costs and benefits associated with education in Northern Nigeria[17] and Argentina.[18] In each of these cases, linear programming techniques based on cost-benefit analysis were used to ask broader questions about the contribution of education to the economy, and about ways of improving the efficiency of educational systems.

It is important to distinguish between the costs and benefits to a

---

[16] Valerien Harvey, "Economic Aspects of Teachers' Salaries," unpublished Ph.D. dissertation, University of Chicago, 1967, p. 35.

[17] Samuel S. Bowles, *op. cit.*

[18] Irma Adelman, "A Linear Programming Model of Educational Planning: A Case Study of Argentina," in *The Theory and Design of Educational Development,* edited by Irma Adelman and Ericke Thorbecke (Baltimore: John Hopkins Press, 1966).

society that invests in education and those which accrue to the individual making an investment in education for himself and his family. Society pays a fairly large proportion of total costs, in our society, at least through secondary school. On the other hand, the entire (pre-tax) income associated with an increment in education contributes to gross national product, while the individual obtains an after-tax benefit from his education. The net result is that social rates of return are lower than private rates of return (Table 2-2).

Table 2-2 Internal Rates of Return to Total and Private Resource Investment in Schooling, United States, Males, 1949 (For Selected Years of Schooling)

Increment in Schooling		Internal Rate of Return to Total Resource Investment (%)	Internal Rate of Return to Private Resource Investment (After Tax) (%)
From Grade	To Grade		
9	10	9.5	12.3
11	12	13.7	17.5
13	14	5.4	5.1
15	16	15.6	16.7

Source: W. Lee Hansen, "Total and Private Rates of Return to Investment in Schooling" *Journal of Political Economy,* LXXI, No. 2 (April, 1963), pp. 128–140.

We turn next to a specific example of private resource investment, namely the investment made by teachers in upgrading their skills. As a basis for the discussion, we refer to a study that was conducted by Valerien Harvey (*op. cit.*), based on data from the Province of Quebec. Using present value and rate of return analysis, Harvey found considerable differences among school districts in the incentive to invest in further education.

Table 2-3 applies cost-benefit analysis to the study of private investment in teachers' education. It is well known that most salary schedules reward teachers for advancing their qualifications. Furthermore, in many cases this incentive appears to be effective, since large numbers of teachers do spend their summers, their sabbaticals, and evenings and weekends in upgrading their qualifications. While they may be doing so in part

Table 2-3 Present Discounted Value at Age 18 of Private Investment in Teacher Training, Adjusted for Pension and Mortality, Using Various Discount Rates; Protestant Teachers. Females and Males, Quebec, 1965 (Thousands of Dollars)*

Discount Rates	Schooling Levels							
	12	13	14	15	16	17	18	19
District 265†								
0.00	294.80	329.58	356.82	448.21	489.65	495.40	509.15	498.75
0.02	164.41	181.03	193.11	235.90	253.34	256.13	260.43	252.09
0.04	103.95	112.43	117.89	139.72	147.16	148.57	148.72	141.92
0.06	72.32	76.74	79.00	90.86	93.74	96.66	92.68	87.06
0.08	53.98	56.18	56.75	63.44	64.09	63.36	61.76	57.05
0.10	42.40	43.31	42.92	46.72	46.23	45.14	43.31	39.33
District 673								
0.00	264.54	292.28	320.96	390.66	409.62	447.41	463.31	453.87
0.02	146.96	161.07	174.44	208.88	217.03	233.87	239.81	232.13
0.04	92.24	100.08	106.60	125.21	128.59	136.37	138.17	131.85
0.06	63.61	62.22	71.35	82.06	83.14	86.64	86.62	81.36
0.08	47.05	49.82	51.14	57.55	57.44	58.78	57.93	53.51
0.10	36.66	38.29	38.57	42.47	41.72	41.92	40.70	36.96
District 061								
0.00	311.61	341.71	370.20	421.54	444.34	435.66	446.97	438.00
0.02	176.38	190.68	203.68	227.11	237.03	229.60	233.64	225.64
0.04	112.92	120.01	126.04	137.21	141.48	135.05	135.34	129.18
0.06	79.33	82.76	85.33	90.60	92.14	86.53	85.49	80.31
0.08	59.64	61.03	61.75	63.94	64.09	59.17	57.56	53.18
0.10	47.10	47.28	46.94	47.44	46.83	42.49	40.69	36.95

* Valerien Harvey, "Economic Aspects of Teachers' Salaries," unpublished Ph.D. dissertation, University of Chicago, 1967, p. 123.

† Done with females' mortality rates, which contributes to slightly overestimate the figures.

because of the consumption value which they attach to education, it may be assumed that they are partly motivated by the increased earnings which they hope to obtain.

Table 2-3 shows that more education is associated with increased present values, at certain discount rates. However, the incentives differ. In District 061, for example, it is not profitable to increase one's schooling from 16 to 17 years even without discounting, since the additional benefits are less than the additional costs. In District 265, it is profitable to increase one's schooling from 13 to 14 years at all discount rates up to 8 percent but not when the discount rate is 10 percent (or more); at rates of 10 percent or more, a District 265 teacher would not find investment in a fourteenth year of schooling economically appealing, though the two-year jump from 13 to 15 years of education would pay off even at a 10 percent discount rate.

In addition to its value in improving private decision making, this type of analysis would be useful to school districts or states that wish to encourage the upgrading of teachers' qualifications. An analysis of costs and benefits would reveal many inconsistencies and incongruities in present salary policy. In some cases, where teachers attend school in the summer and evenings (and thus have a low level of foregone earnings), the profitability of increased training is very high; in other cases, there may not be sufficient incentive to encourage teachers to take additional work. In all cases, the rate of discount that is used in the model is a critical factor.

We turn now to the other analytic tool—rate-of-return analysis. In the case of teachers, the rate of return to increased education varies among subcategories. The following data (Tables 2-4 and 2-5) show differences among school districts in Quebec in the profitability of investment in additional education. Table 2-4, for Catholic teachers, shows one school district in which education is a very desirable investment, one in which continuing education is a more dubious investment, and one school district in which continuing education is even penalized at certain levels.

Table 2-5, for Protestant school districts, shows a more even pattern, although continuing education is still rewarded more strongly in some cases than in others.

These examples deal primarily with private decisions, but they also have implications for public agencies. Salary schedules may be used as an incentive to persuade individuals to upgrade their qualifications. Whether by design or accident, salary schedules may affect teacher mobility, since a movement from one school district to another may often be the best way for a teacher to improve his economic status.

Table 2-4 Internal Rates of Return to Private Resource Investment in Teacher Training, Adjusted for Pension and Mortality; Catholic Teachers, Males in Elementary Schools, 1965*

Lower School	\multicolumn{7}{c}{Higher Schooling Levels}
	13	14	15	16	17	18	19
\multicolumn{8}{c}{District 645}
12	7.41	8.21	8.29	9.40	9.62	9.06	7.56
13		8.93	8.69	10.00	10.10	9.35	7.58
14			8.48	10.54	10.48	9.46	7.31
15				12.44	11.42	9.78	7.04
16					10.40	8.38	5.07
17						6.09	1.73
18							−10.87
\multicolumn{8}{c}{District 747}
12	(−8.51)	.38	2.43	4.39	5.32	5.80	6.03
13		3.63	4.79	6.50	7.11	7.32	7.35
14			5.96	7.89	8.24	8.23	8.08
15				9.79	9.37	8.98	8.61
16					8.98	8.59	8.21
17						8.22	7.84
18							7.49
\multicolumn{8}{c}{District 542}
12	2.93	2.81	2.70	1.32	.40	− .29	− .84
13		2.72	2.59	.74	− .36	− 1.12	− 1.71
14			2.50	− .41	− 1.72	− 2.57	− 3.21
15				− 9.11	−10.30	−11.02	−11.57
16					− 9.25	−10.45	−11.20
17						− 9.39	−10.62
18							− 9.55

*Valerien Harvey, "Economic Aspects of Teachers' Salaries," unpublished Ph.D. dissertation, University of Chicago, 1967, p. 113.

Note: The negative rates of return are a result of having salary schedules such that salary differentials resulting from more education add up over a lifetime to less than the cost of the increment in education.

Table 2-5 Internal Rates of Return to Private Resource Investment in Teacher Training, Streams Adjusted for Pension and Mortality; Protestant School Teachers, Quebec, 1965*

Lower Schooling Levels to from	\multicolumn{7}{c}{Higher Schooling Levels}
	13	14	15	16	17	18	19

				District 673 Montreal			
12	16.12	13.10	15.49	13.66	13.01	12.01	10.14
13		10.79	15.27	13.03	12.41	11.35	9.34
14			19.55	14.16	12.97	11.51	9.04
15				7.78	9.24	8.28	5.73
16					10.54	8.53	5.09
17						5.99	1.38
18							−10.96

				District 265			
12	12.33	10.67	13.19	12.11	11.30	10.38	8.85
13		9.08	13.57	12.05	11.06	10.00	8.27
14			17.55	13.36	11.71	10.25	8.12
15				9.00	7.94	6.88	4.56
16					5.84	5.03	1.64
17						4.57	.66
18							−11.68

				District (061)			
12	10.43	9.83	10.25	9.86	7.84	7.36	6.19
13		9.26	10.17	9.68	7.19	6.72	5.43
14			10.93	9.89	6.50	6.06	4.59
15				8.32	2.86	3.34	1.74
16					−10.23	.58	−1.01
17						4.33	.54
18							−10.65

*Valerien Harvey, "Economic Aspects of Teachers' Salaries," unpublished Ph.D. dissertation, University of Chicago, 1967, p. 122.

Summary

This chapter has used open systems theory in the development of models depicting the nature of the resource exchange between schools and their environments. It has closed by describing some research which has already been conducted. However, the hoped for goal of this chapter is that it will stimulate additional empirical study.

Two steps remain before an attempt is made to suggest the implications of this analysis for decision making. The first step, to be undertaken in Chapter 3, is to outline procedures for studying educational costs. The second step is to suggest more refined analytic procedures for discussing the relationships among inputs as well as between inputs and outputs; this is the thrust of Chapter 4.

Chapter Three

❖❖❖❖❖❖❖❖❖

The Analysis of Costs

One of the most important contributions of economics to educational administration is a general concept of costs. Such a concept has important implications for a conceptual approach to decision-making. It is also a valuable tool for the practicing administrator.[1]

The usual practice in education is to regard costs as synonymous with expenditures. Costs are taken to include the monetary outlay associated with the purchase of factor inputs, such as teachers' and administrators' salaries, books and materials, equipment, buildings, land, and school buses. In the following discussion, we include these items as costs but include as well some nonexpenditure items encompassed in the concept of "opportunity costs." In fact, all costs may be defined in terms of "what is given up" rather than "what is put in."[2] This more general concept of costs has a number of important advantages.

In the first place, the opportunity cost concept focuses attention on the importance of students' time. When individuals who would otherwise

---

[1] This discussion relies heavily on Mary Jean Bowman's "The Costing of Human Resource Development," in *The Economics of Education,* edited by E. A. G. Robinson and J. E. Vaizey (New York: St. Martin's Press, 1966), pp. 421–450.

[2] *Ibid.,* p. 425. Opportunity costs are defined as follows: "The opportunity costs of choosing a commodity, service, or activity 'A' . . . are what the individual or group or society gives up . . . in making this choice" (*Ibid.,* pp. 422–423).

be in the labor force are in school or college, their productive services are withdrawn from the economy, and they themselves forego earnings. Schultz has presented data suggesting that 60 percent of school costs in high school and college consist of earnings foregone (Table 3-1).

Table 3-1 School Costs, Earnings Foregone, and Total Costs of Schooling per Student per Year in the United States*

United States, 1956 (Dollars)	School Costs	Earnings Foregone	Total	Earnings Foregone as Percentage of Total Costs
8 years elementary	280	0	280	0
4 years high school	568	852	1420	60
4 years college or university	1353	1947	3300	59

* Theodore W. Schultz, *The Economic Value of Education* (New York: Columbia University Press, 1963), p. 29.

Internal time allocation might better be governed by the principle of "foregone learning"[3] rather than "foregone earning." The implication here is that the cost of a given curriculum or of a given instructional procedure is measured in part by foregone opportunities to devote students' and teachers' time to other curricula and procedures. Thus, part of a student's cost in attending a class in biology consists of foregoing his opportunities to use this time studying physics or literature. Part of the cost of attending classroom lectures consists of the foregone opportunity to spend this time reading in the library, or engaging in other forms of self-instruction. This concept is a very important one, and has special implications for the scheduling of pupils. It will be discussed in more detail in Chapter 6.

There are additional implications for the practicing administrator. When the administrator and the teacher recognize that students' time is a valuable resource, they will attempt to use it to the best possible advantage. The concept can be generalized throughout the educational system. For example, when a teacher's services have been purchased, there are opportunity costs associated with assigning her to any specific activity. These costs consist of alternative activities which must be foregone. A decision to assign a teacher to lunchroom supervision is not

[3] The concept of "foregone learning" was suggested by Mary Jean Bowman in informal discussion.

costless; among other possibilities, the system foregoes her presence in a classroom. Thus opportunity cost analysis may suggest that lower paid persons should be used for nonteaching duties. Similar analysis may be applied by the administrator to the use of his own time. I leave to the reader a further exploration of this application of the concept of opportunity cost.

A second application of the opportunity cost concept is embedded in the system of incentives which are used to encourage teachers and administrators to upgrade their qualifications.[4] Educational administrators who decide in the middle of their career to return to graduate school to complete their doctorate must weigh both direct and indirect (opportunity) costs against the increased income they expect to receive. They may decide to enter a graduate school that permits them to complete course work and dissertation requirements in the evenings and during the summer, thereby greatly increasing the rate of return to their investment by reducing the foregone earnings element in their decision. In industry, opportunity costs are a factor in the decisions of companies to provide (and of individuals to profit by) on-the-job training opportunities, since when employees are attending a course they are removed from productive work for at least part of their time.[5]

These examples suggest that the concept of cost is much broader than commonly recognized. Every decision of the administrator, whether or not it involves the expenditure of money, has costs as well as benefits associated with it. Because of the ubiquity of costs, we proceed to analyze the concept in more detail.

THE CATEGORIES OF EDUCATIONAL COSTS

1. Direct and Indirect Costs

In a broad sense, all costs, as noted above, are opportunity costs. However, a distinction must be made between direct costs, which entail

---

[4] Research bearing on the relationship between teachers' salary schedules and upgrading was reported in Chapter 2. Opportunity costs have an important bearing on teachers' decisions to upgrade their qualifications.

[5] Gary S. Becker, "Investment in Human Capital: A Theoretical Analysis," and Jacob Mincer, "On-the-Job Training: Costs, Returns, and Some Implications," in *The Journal of Political Economy,* **LXX,** No. 5, Part 2 (Supplement, October, 1962), pp. 9–49 and 50–79. Also, Gary S. Becker, *Human Capital* (New York: Columbia University Press, 1964).

a monetary outlay, and indirect costs, which do not entail an expenditure of money, and must be valued in terms of foregone opportunities, including the loss of earnings.

Direct costs involve hiring teachers, administrators, counsellors, and janitors, and purchasing equipment, materials, land and buildings. Direct costs or expenditures are usually included in the educational budget, and reported to the public and appropriate legislative bodies.

Indirect costs include depreciation and the obsolescence of buildings, as well as imputed interest on invested capital. They also include property and sales tax exemptions granted to educational institutions. Most important, indirect costs include students' foregone earnings. Rational decision making demands that these costs be taken into consideration, since they involve important foregone opportunities for individuals and society.

2. Social and Private Costs

One important variable in the provision of education is the degree to which the cost falls upon the individual and his family, instead of on the total society. Private costs are borne by household units; social costs include private costs and in addition, the costs which are shared among members of the wider society.[6]

Private costs represent foregone opportunities for individuals and their families. Parents who invest their earnings in their children's education rather than in housing or in the stock market may well evaluate the costs in terms of the opportunities foregone. The student entering college may reflect on a number of private costs—the money spent for tuition and other fees, the earnings he will forego, and the loss of leisure.

Private costs are reduced when there is a significant public contribution to the support of schools and colleges. Society's support of at least part of the total costs is made possible through money collected by taxation, gifts, or bequests. In elementary school, public monies approach the total outlay of resources; in private universities, the student pays a considerable fraction of the total direct cost, although these institutions also receive public subsidies.

---

[6] Costs and benefits which accrue to people other than students and employees of educational organizations are called "external" effects or externalities. Friedman calls them "neighborhood effects" [Milton Friedman, "The Role of Government in Education," in *Capitalism and Freedom* (Chicago: University of Chicago Press, 1962)]. The sum of private and public effects is equal to the social costs and benefits of education.

Social costs are the sum of private costs and costs paid by the public. Total direct and indirect costs may be compared with social benefits, to provide a basis for decisions concerning social investment in education.

3. Monetary and Nonmonetary Costs

Monetary costs may be either direct or indirect, and may be paid by either society or the individual. The payment of tuition fees involves a direct monetary cost to the student and his family, while foregone earnings constitute an indirect monetary cost. Nonmonetary costs include opportunities to enjoy leisure which are foregone by the student when he spends long hours with his books. From a social point of view, nonmonetary costs include the discomfort incurred by those who are disturbed by noisy playgrounds in the vicinity of a school.

Within the school, nonmonetary costs include the opportunity costs associated with the allocation of time. Each assignment—of a student to a room, with a given teacher, to study a given subject—carries with it the costs of foregone learning in other subjects and in other situations. These types of nonmonetary costs constitute a major portion of the total costs of operating a school plant. They are discussed in more detail in Chapter 6.

It is now possible to turn to the problems involved in analyzing costs, both at the national and state level, and in individual schools. We begin with an examination of the problems associated with aggregate or macro cost analysis.

AGGREGATE COST ANALYSIS

Both economists and educators have examined total national costs of education. Economists have been interested in the relationship between the cost of education and economic growth. Their questions include the following:

1. What fraction of total resources of the nation has been put into education?
2. How big is the education component of national income at factor prices?
3. What has education contributed to growth in national income?

4. What was or would be the cost of devoting such and such resources to education?[7]

These questions are important in determining national policy regarding education. The fourth question, which involves cost elements more specifically than the first three, deals with alternatives such as are implied in the question, "What was or would be the cost of devoting such and such resources to education instead of to other things?"[8]

The statistics used by educators have been addressed to a different set of questions. In particular, educators have been engrossed in the problem of securing sufficient resources to operate the schools. However, decisions probably take indirect costs into consideration, at least implicitly. An example of aggregate analysis which considers some types of indirect costs is now presented. The ingredients of cost are first described.

THE INGREDIENTS OF EDUCATIONAL COSTS[9]

Included in the economic costs of education are, first, the types of cost for which statistics are kept by the U. S. Office of Education. These include instructional and administrative costs, and the cost of operating and maintaining the school buildings.

Second, there are the indirect costs of the depreciation of educational facilities. While common practice usually ignores this item, depreciation does constitute part of the total cost of operating an educational system. In the examples appearing later in this section, the cost of depreciation is imputed as one element in total cost.

Third, interest on capital outlay should be included. Educational accounting includes interest actually paid on outstanding bonds as a cost. However, the public's equity in the "paid up" portion of the investment could, if invested elsewhere, draw interest. The foregone interest is part of the cost of owning capital, and should, for purposes of aggregate accounting, be included as a cost item.

---

[7] Bowman, *op. cit.*, p. 430.

[8] *Ibid.*, p. 430.

[9] This analysis was patterned after that in Rudolph C. Blitz, "The Nation's Educational Outlay," in *Economics of Higher Education,* edited by Selma J. Mushkin (Washington, D.C.: U. S. Department of Health, Education, and Welfare, Office of Education, 1962), pp. 147–169.

The fourth element in total costs is exemptions from the property taxes and sales taxes granted to educational institutions. These exemptions may be viewed as public subsidies to educational systems. They are costs to the public sector, because the public gives up this potential income. To the extent that the subsidy is a hidden one, seldom recognized and even more seldom discussed, it can result in a distortion of the decision-making process.

The Michigan Study

In a recent financial study in the year 1965–1966 in the state of Michigan an attempt was made to estimate the total direct costs of education.[10] The following items were included (Tables 3-2 and 3-3):

1. Operating Costs, Public Elementary and Secondary. This item was the same as that reported in state and national expenditure statistics.
2. Operating Costs, Nonpublic Elementary and Secondary. These figures were based on the assumption that per student costs, including contributed services, are the same in the public and nonpublic sectors. It was further assumed that, of the cost of operating nonpublic schools, 75 percent was in the form of direct monetary outlays, and 25 percent was in the form of contributed services of faculty.
3. Imputed value of depreciation and interest. The procedure used was that reported by Rudolph C. Blitz.[11] Arbitrarily, again, per student costs for this item in nonpublic elementary and secondary schools were assumed to be equivalent to those in the public sector.
4. Imputed value of property tax exemption. Since the statewide ratio of state equalized value to full value in Michigan was about 50 percent, half of the figures for building and site value were taken as the property on which taxes are foregone. (The assumption was that school buildings replace other buildings of equal value. This was probably a poor assumption in residential and rural areas, because school buildings here are probably of greater value than the property they replace; in industrial and urban areas their value is probably often less.) It was estimated that the statewide average property tax in 1965–1966 was 36.8

[10] J. Alan Thomas, *School Finance and Educational Opportunity in Michigan* (Lansing: State Department of Education, 1968).
[11] Rudolph C. Blitz, *op. cit.*, p. 160.

mills on state equalized valuation, assuming a total equalized valuation in that year of about $28.2 billion. Again, nonpublic schools were assigned a prorated value.

INDIRECT COSTS

The value of students' time was not included in the Michigan calculations. This is an important variable for *national* income accounting. It is also a critical element of *private* cost. Furthermore, as shown above,

Table 3-2 Michigan's Annual Educational Costs, Elementary and Secondary, Exclusive of Students' Foregone Earnings, 1965–1966

Direct Costs		Amount
1. Operating costs, public elementary and secondary		$ 918,943,000
2. Operating costs, nonpublic elementary and secondary	$121,624,000	
Contributed services (est.)	40,541,000	
Total		162,165,000
3. Imputed value of depreciation and interest public elementary and secondary		244,296,000
Nonpublic elementary and secondary		43,111,000
4. Imputed value of property tax exemption, public elementary and secondary		51,210,000
Nonpublic elementary and secondary		9,037,000
Total		$1,428,762,000

Note: The following procedures were used in calculating the distribution of physical assets of public elementary and secondary schools. It was assumed that "20 percent of their assets are in land, 72 per cent in buildings, and 8 percent in equipment. There is no depreciation in the value of land. The depreciation of buildings is calculated at 3 percent per year. Although buildings are assumed to have a lifespan of 50 years, a period that would warrant only 2 percent depreciation, 3 percent is imputed to that value to make some allowance for obsolescence due to population shifts. Ten percent depreciation is imputed to equipment; to this is added implicit interest of 5 percent—a total of 8 percent for depreciation and implicit interest." From Rudolph C. Blitz, "The Nation's Educational Outlay," in *Economics of Higher Education,* edited by Selma J. Mushkin (Washington, D. C.: U. S. Department of Health, Education, and Welfare, Office of Education, 1962) p. 160.

(The data used in these calculations were obtained from reports made to the State Department of Education on Form A statistical reports. There is little doubt that local school districts show discrepancies in their interpretation of the term "total value." It is not clear, for example, whether initial cost or replacement cost is desired by the state authorities for this category.)

students' time as a cost affects decisions of educational administrators responsible for scheduling. However, from the point of view of the *state* decision maker, this ingredient of cost is of limited importance.

Table 3-3 Total Value of School District Assets in the Form of Building, Site and Equipment 1965–1966

Building value	$2,603,402,406
Site value	179,773,407
Equipment value	210,529,257
Total	$3,053,705,070

MICROANALYSIS OF COSTS

There are important differences between the interests of educators in cost analysis and those of economists. Economists have tended, until now, to emphasize the broader questions of social policy discussed above. Educators, on the other hand, are increasingly interested in costs as they affect decision making. However, economists are beginning to turn their attention to questions about the production of education.[12]

Purpose of Cost Analysis

1. At one level, cost data are primarily descriptive. Information about the cost of building and operating a school, of providing a course in vocational agriculture, or of providing special training for the chronically unemployed is important for comparative and historical purposes.

[12] See, for example, Henry M. Levin, "Cost-Effectiveness Analysis and Educational Policy—Profusion, Confusion, Promise," *Research and Development Memorandum No. 41,* Stanford Center for Research and Development in Teaching, School of Education, Stanford University. Also, Samuel Bowles, "Towards an Educational Production Function," paper presented at the Conference of the National Bureau of Economic Research on "Education and Income," Madison, Wisconsin, November 15–16, 1968.

2. More important, cost data reveal *what is being given up* in terms of other alternatives for which a given set of resources might be used. For example, an item of two million dollars for the operation of a school building may be considered to be a measure of the instructional improvements which are foregone.

Cost analysis is therefore an essential element in internal decision making, since a decision or choice from among alternatives means giving up certain options; costs may be regarded as measures of what is foregone or given up. This does not imply that decisions may be made on the basis of costs alone. The benefits as well as the costs of the various alternatives that are considered comprise, in both intuitive and rigorous terms, the proper bases for making decisions.

3. The careful analysis and reporting of cost data provides a means of *control* over the internal operation of educational systems. Such control is necessary to ensure wise and proper use of funds.
4. Finally, unit cost data provide an important input into certain kinds of research. Operational research, which may be defined as the scientific study of management, requires cost data.

One application of such research would involve a decision as to whether two or more small rural high schools should be consolidated into a larger school. Additional costs of transporting students may be balanced, to some extent, by unit cost savings associated with the use of larger classes, reduced administrative overhead, and other economies. The final decision will take into consideration political pressures, sociological concerns, and educational efficiency. However, cost items cannot be ignored, and must be weighed against the net nonmonetary costs and benefits.

The manner in which costs are measured and reported will depend in large part on the purpose of the analysis. This in turn will be partly determined by the organizational unit which is being considered, and by the type of decision which is being made. The next section examines some of these variables.

THE ORGANIZATIONAL CONTEXT OF COST ANALYSIS

Let us first consider the question of organizational context. *It may be stated as a general principle that the lower one is in the organizational hierarchy, the more his decisions will involve allocating time rather than money.*

1. At the school district level, there will be a concern for the costs associated with the purchase of goods and services. As part of their responsibility for operating an efficient educational system, school board members and the school superintendent may study the monetary costs of operating a given school, or of providing a given type of program. There is typically much less concern with the manner in which the input, once purchased, is used. This type of problem is left, in part at least, to the school principal.
2. To the extent that each school principal has control over the budget of his school, some costs at the school level will be expressed in monetary terms. However, many school principals do not have their own budget, and the major resource under their control is the time of teachers and students. Decisions concerning the deployment of personnel involve costs, which must be evaluated in terms of foregone opportunities to use this time in other ways.
3. Within each classroom, the teacher is a decision maker who determines, among other things, how students' time and her own time as well shall be used. Teachers sometimes have control over petty amounts of cash, including some money raised outside the regular budget, but the major costs that teachers take into consideration in their decision making are foregone opportunities to use time in other ways.
4. Students also deal with opportunity costs, when they are given some discretion over the way in which they spend their time. Some secondary schools and all universities force a kind of cost analysis on students by requiring them to decide how they will divide their time among the various subjects on the curriculum. Time is by far the most important resource which is subject to the control of students.

In each of these contexts, it is assumed that resources should be allocated so as to maximize the achievement of a given objective or set of objectives. Cost analysis provides a means for examining the manner in which resources are used. The data may therefore give evidence concerning the rationality of the organization. However, monetary costs are not always subject to administrative control and are not always identical with foregone opportunities. Variations in prices of inputs, including teachers' salaries, may affect costs. Costs may also be determined in part by the amount of money which is available. For example, teacher salary agreements are often affected by the level of financial support provided by the state legislature. Finally, the allocation of resources to meet a given set of objectives may be an ambiguous criterion of rationality, since the objectives themselves may be partially determined by the amount of resources available. At any rate, it is important to develop procedures for cost analysis in education. Some aspects of these procedures are described below.

Elements of Unit Cost

Cost differences may result from many factors, including the quality of services provided and the price paid for inputs. Also, they may be due to differences in the mix of outputs provided by educational systems. Systems with a more expensive output mix may for this reason show high costs.

Table 3-4 illustrates how differences in the output mix may affect the total cost of operating a school. The table compares two hypothetical schools of 1000 students each. Students in each school register for 5 courses each, for a total of 180 hours in each course.

The differences in total cost between the two schools are due to differences in output mix. School B has more students enrolled in the more expensive courses (woodworking, homemaking, and business),

Table 3-4 Differences in Output Mix in Two Hypothetical Schools (Assume Identical Cost per Student Hour in Schools A and B)

Subject	Course Enrollment School A	Course Enrollment School B	Cost per Student Hour	Total Cost School A	Total Cost School B
English	1000	1000	0.35	$ 63,000	$ 63,000
Mathematics	600	400	0.35	37,800	25,200
Phys. Educ.	1000	1000	0.50	90,000	90,000
French	500	200	0.35	31,500	12,600
Woodworking	200	400	0.60	21,600	43,200
Shop	100	400	0.50	9,000	36,000
Science	500	300	0.45	40,500	24,300
Business	300	500	0.40	21,600	36,000
Soc. Studies	800	800	0.35	50,400	50,400
Total	5000	5000		$365,400	$380,700

while school A has most of its students enrolled in the less expensive, academic types of courses. It would therefore be misleading to compare the per pupil costs in the two schools ($365.40 as opposed to $380.70) without including in the comparison a statement concerning the differences in the product mix of the two schools.

Also, expenditure comparisons are based on very gross units of time. School years and school days vary in length from system to system; therefore a measure of costs on an annual and daily basis would be misleading. Comparative data should be based on comparative time units. The best unit is therefore the cost per student hour for a specific service.

In determining the direct cost of a student hour of a given service, the following element must be considered.

Teacher's salary
Other salaries
Space
Equipment and materials

On the basis of these factors, we present the following example of unit cost analysis.

Consider the components of the cost per student hour of a course in Biology I. We must include the costs of the time of teachers and other personnel, of space, equipment, and books. We will assume that the knowledge of subject matter upon which the course is based does not represent a cost to the school. The school, however, must bear the costs of books and of the services of a teacher who has mastered the knowledge which the course contains.

1. Teacher's time. If there are other people such as the chairman of the science department or science consultants who share the task of providing instruction in biology, we must include the costs of their time. If the teachers' task is defined as including preparation for classes, grading paper, and setting up experiments, this time should also be charged against the teacher's inputs.

Assume that the teacher's salary is $8000, and that she is assisted by supporting personnel to the extent of another $1000 in salary. Assume that this teacher teaches a total of 120 students, each for 180 hours in the school year. Then the cost per student hour for teachers' services in Biology I is $0.42.

2. Administrator's time. One important administrative function is to improve the quality of educational services. Hence, part of the time of administrators should be prorated among the subject matter areas. We will assume, in our example, that the costs of providing clerical assistance and administrators' time are included in the time of supporting personnel reported in *1*.

3. *Space.*[13] Scheduling is one of the central roles of the administrator, since it is directly related to facilitating teaching and learning, and since it involves the allocation of key resources—time of students, time of teachers, space and equipment. The administrator must have some knowledge of the relative cost of resources if he is to use them productively. However, educators often pay insufficient attention to the cost of space. The concern in public administration generally is for out-of-pocket expenditures rather than for the economic cost of resource inputs.

There are five components to the cost of classroom space. The first is the interest on unpaid debt. The second is the economic cost of the equity, or the interest this equity would produce if it were invested elsewhere. The third is depreciation—the annual decrease in value due to the aging of the building. Fourth is the overhead associated with space—heat, light, and power. Fifth is the cost of maintenance. For comparative purposes, consider the cost of home ownership to a family who owns a house which originally cost $20,000.00. Assume that at a given point in time, the family has a mortgage for $8,000.00 but that, due to depreciation, the value of the house is now $16,000.00. The cost of home ownership includes the following components:

(a) Interest on the mortgage. Say 4% of $8000	$ 320
(b) Imputed interest on the equity. Say 5% of $8000	400
This is the amount which the equity would bring if it were invested elsewhere—assuming a going rate of interest of 5%.	
(c) Depreciation. Assume an additional expected life of 25 years ($16,000/25)	640
(d) Taxes. Say	640
(e) Maintenance. Say	400
(f) Overhead (light, heat, power). Say	600
Total cost of space	$3000

Now let us cost out the space of the laboratory facilities for Biology I. Assume a biology laboratory containing 1200 square feet. Assume

---

[13] *Note:* This discussion ignores costs due to obsolescence. These costs also are considerable in education. The rate of obsolescence of school building depends in part on the degree to which there is a recognition of possible future needs for change at the time of initial construction of a school building. For example, flexibility may be provided and obsolescence may be retarded by using inner walls which can be readily removed or altered.

another 25 percent of this space (300 square feet) for supporting space (corridors, etc.). Let the cost of construction be $16 per square foot, making a total cost of $24,000. Assume further that the present value (after depreciation) is $20,000, of which $15,000 is still owing. The cost of this space may be calculated as follows:

(a) Interest on debentures. Say 4% of $15,000	$ 600
(b) Imputed interest on equity. Say 5% of $5000	250
(c) Depreciation. Assume additional expected life of 20 years	1000
(d) Maintenance. Say	400
(e) Overhead (light, power, heat). Say	400
Total cost	$2650

This cost is for, say 120 students for 180 hours each, making a per student-hour cost of $0.12.

4. Equipment. Assume that the biology laboratory is equipped to the value of $10,000, all of which has been paid. The cost can be calculated as follows:

(a) Imputed interest 5% of $10,000	$ 500
(b) Maintenance. Say	100
(c) Depreciation. Assume additional expected life of 10 years.	1000
Total	$1600

which comes to $0.06 per student hours.

5. Materials, totalling say, $0.02 per student hours. Thus, to recapitulate, the cost per student hour is as follows:

Personnel	$0.42
Equipment	0.06
Space	0.12
Material	0.02
Total	$0.62

ENROLLMENT AND THE COST OF EDUCATIONAL SERVICES

In this section, we will apply our knowledge of cost accounting to a discussion of the relationship between enrollment and the cost of providing a given service (say, Biology I). It is possible to proceed from this analysis to a discussion of the effects of scale on the costs of varying mixes of services. The advantage of this procedure is that it separates the effect on cost of the nature of the service mix from the effect on cost of the scale of operation.

We are dealing, as before, with individual secondary schools, rather than with school districts or with state or national school systems. The importance of this distinction is that it enables us to consider input prices as "givens." The study ignores the external effects of scale, which include the effect on input price of increases in the magnitude of educational operations, and also the possible growth of complementary industries (such as educational TV) as a result of growth in the educational sector of the economy.

At the higher levels of organization in education, say the state school system, increases in the demand for teachers, school buildings, equipment, and books may cause the prices of these inputs to rise. For example, if the demand for teachers increases more rapidly than the number of teachers who are available, districts will bid against each other for teachers' services. As a result, teachers can obtain higher salaries, and the supply prices of teachers' services increases.

In the case of the individual school, however, an increase in size does not result in a sufficiently larger demand for the services of teachers and for material inputs to affect the supply price. Of course, it is possible that a school may increase in size concurrently with an increased total demand for factor inputs. In this case, the rise in price due to total increase in demand must be separated from the relationship between the size of the school and the cost of the programs which it offers.

We now turn to a more detailed analysis of the effect of the size of a school on costs (Table 3-5). Returning to our example of the case of Biology I, assume the following. (1) As additional teachers are added, the $1000 for additional personnel does not change. (This means that the consultative and other services are adequate for use with additional teachers.) (2) As additional classrooms are added, the amount of additional equipment that is necessary is $6000 per class instead of $10,000. (This means that some expensive equipment can be used for more than one class). This assumption seems reasonable because the more expensive items of equipment are usually not in constant use.

We will assume that biology laboratories are built to serve 120 students. (This is 24 students per period for a five period day.) However, we assume further that these rooms can be overloaded (by increasing class size and by increasing the day to a total of 7 hours) to a maximum capacity of 200 students. At this point, an additional laboratory will be added.

Finally, we will refer to personnel, space, and equipment as indivisible units, and we assume that the number of units will be determined on the basis of enrollment intervals. On the other hand, there will also

Table 3-5 Cost Schedule as Additional Units Are Added

1. One teacher, one classroom
    - Personnel — $9,000
    - Space — 2,650
    - Equipment — 1,600
    - Total — $13,250

2. Two teachers, two classrooms
    - Personnel: $9,000 / 8,000 — 17,000
    - Space: 2,650 / 2,650 — 5,300
    - Equipment: 1,600 / 950 — 2,550
    - Total — 24,850

3. Three teachers, three classrooms
    - Personnel: $9,000 / 8,000 / 8,000 — 25,000
    - Space 3 × $2,650 — 7,950
    - Equipment: 1,600 / 950 / 950 — 3,500
    - Total — $36,450

be variable costs (for supplies and materials) which, we assume, will total $10 per student.

The first column of Table 3-6 shows fixed costs, assuming that additional units (space, equipment, and teachers) are added for each 200 students. Column 2 shows the amount of variable costs. Column 4 shows the total cost for the school per hour of instruction in Biology I. Column 5 is obtained by dividing Column 4 by the maximum enrollment in that interval. Column 6, the average student cost per hour is obtained by dividing Column 5 by 180, assuming that each student has 180 hours per year of instruction in Biology I. Finally, Column 7 presents the incremental costs per student per hour. This is the increase in total costs from one interval to another divided by 50 (the size of the enrollment interval) and by 180 (the number of student hours in the year). It is clear that incremental costs increase substantially when new units are added, but that the increase in incremental costs is very small when they involve only an increase in the supplies which are consumed.

Figure 3-1 graphs the changes in average cost which are associated with increased size. It is clear from Figure 3-1 that economies of scale

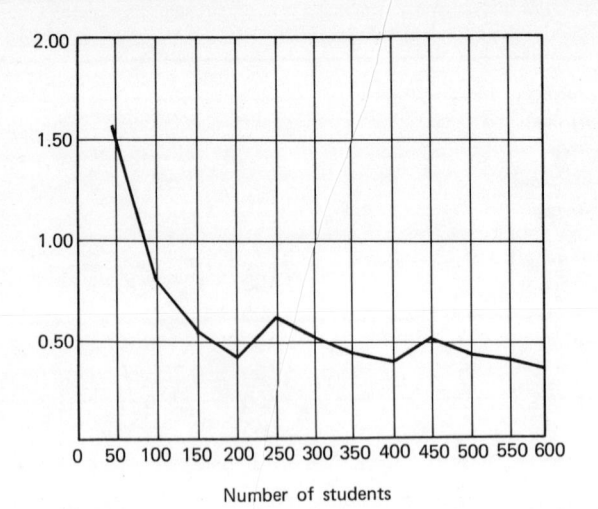

Figure 3-1 Average cost per student per hour of instruction in Biology I.

(associated with a reduction in cost per student hour as school size increases) are quite substantial in our example. Let us now return to the factors which, in our example, are the cause of these economies.

Economy of scale was built into the analysis in the form of the improved utilization of personnel other than teachers. It will be recalled that we allowed $1000 for supplementary salaries, and that we left this figure unchanged as additional units of space and equipment were added. This is meant to reflect the idea that the services of decision-making personnel (administrators, department heads, supervisors, and so forth), can be shared among additional individuals. However, like teachers and classrooms, additional units of these personnel might be required as the size of the total operation increases. When that point arrives, it is possible that the economies may be obtained through greater specialization and division of labor.

Finally, the economies of scale reflect the fact that space and equipment are used more intensively in a larger operation than in a smaller one. Again, it is possible that these savings may go into the purchase of additional items of equipment, rather than into cost reduction.

It must, however, be emphasized that there probably is a point beyond which economies of scale will no longer be possible. In fact, the graph in Figure 3-1, if continued, might begin to show an increase in average costs (Figure 3-2). The reason for this phenomenon is that, as size increases beyond the optimum point, the organization may become more difficult to administer. Coordination of the activities of teachers

Table 3-6 Fixed and Variable Costs For Given Enrollment Intervals (Total and Per Student)

Enrollment in Biology I	"Fixed" Costs (For Teachers, Space, Equipment)	"Variable" Supplies, $10 Per Student	Total	Total Cost Per Hour (Assume 900 Hours)	Average Cost Per Student Per Year	Average Cost Per Student Per Hour (Assume 180 Hours in a School Year)	Incremental Cost Per Student Hour
50	13,250	500	13,750	15.28	275.00	1.53	1.53
100	13,250	1,000	14,250	15.83	142.50	0.79	0.05
150	13,250	1,500	14,750	16.39	98.33	0.55	0.05
200	13,250	2,000	15,250	16.94	76.25	0.42	0.05
250	24,850	2,500	27,350	30.39	109.40	0.61	1.34
300	24,850	3,000	27,850	30.94	92.83	0.51	0.05
350	24,850	3,500	28,350	31.50	81.00	0.45	0.05
400	24,850	4,000	28,850	32.05	72.12	0.40	0.05
450	36,450	4,500	40,950	45.50	91.00	0.51	1.34
500	36,450	5,000	41,450	46.05	82.90	0.46	0.05
550	36,450	5,500	41,950	46.61	76.30	0.42	0.05
600	36,450	6,000	42,450	47.17	70.75	0.39	0.05
	(1)	(2)	(3)	(4)	(5)	(6)	(7)

and students may, for example, require the services of additional personnel. It will be necessary to devise procedures to prevent excessive depersonalization within the organization, which might be counterproductive on an educational organization. However, we do not as yet know what this optimal point may be, and it will vary from one school to another

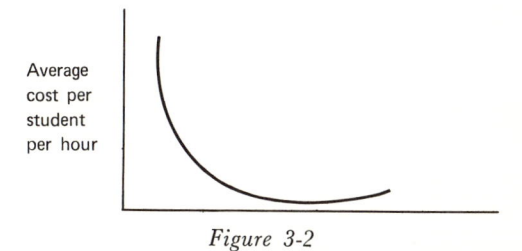

Figure 3-2

depending on such factors as (a) population density; (b) homogeneity or heterogeneity of the student body; (c) availability of arteries of transportation; (d) organizational know-how.

VARIATIONS IN OUTPUT MIX

In the above discussion we have shown how the input-output relationship can be measured, in terms of cost per student hour of a given course. We have also pointed out the factors which may lead to economies as the size of the school increases. In this section, we turn to the problems which are involved when programs are combined to form a comprehensive educational curriculum.

We will refer to the aggregation of programs as an "output mix." We can discuss the output mix for a school, or for an individual student who studies, say, five courses, for an hour each, every day. Since the costs per student-hour vary among programs, they will also vary from school to school, in cases where there are differences in enrollments among subject areas. Referring back to the hypothetical costs of Table 3-4, we now compare costs per day for two students (Table 3-7).

Our discussion of the relationship between size of school and cost per student hour in a given program must now be broadened. As the size of schools increases, schools tend to offer a greater variety of courses. There are several reasons for this tendency. In the first place, it requires a relatively large total number of students to produce enough individuals who are interested in and qualified for such courses as a second or a third language, advanced placement physics, or electronics, to make such a course possible. In the second place, large size is usually, although

Table 3-7 Hypothetical Costs per Day for Two Students

Student A's Program	Cost per Hour	Student B's Program	Cost per Hour
English	$0.35	English	$0.35
Mathematics	0.35	Social Studies	0.35
Social Studies	0.35	Physical Education	0.50
Physical Education	0.50	Homemaking	0.50
French	0.35	Business	0.40
Cost per five hour day	$1.90		$2.10

Note: Costs will also be higher when a student enrolls for more than five courses. Hence, within a given school, we may expect input variations among students as the output mix changes.

not necessarily, accompanied by heterogeneity in the total student body, so that a more diversified curriculum is called for.

We are now in a position to present a proposition which is testable, and which provides one means of studying the effect of the size of a school on the economics of its operation. The proposition is as follows:

1. As the size of a school increases, certain economies in terms of the cost per hour of a given program are made possible. These economies may be studied by the analytic means outlined above. They are related, primarily, to (a) indivisibilities; and (b) specialization and division of labor.
2. As the size of a school increases, there is a tendency for the output mix to change in the following ways. (a) It becomes broader, encompassing a greater total number of programs. (b) It includes a greater emphasis on the more costly courses, which require either expensive equipment and space, or the services of highly trained specialist teachers.
3. On an aggregate basis, changes in the output mix tend to obscure the economies of scale. In other words, these economies are used to finance a more diversified program, which includes greater enrollments in more costly program areas.[14]

[14] Bowser, *op. cit.* Bowser concludes, in part, as follows (pp. 146–147):

As diversification increases, the cost of the courses added to the service mix in general are higher than the cost of the existing courses. As school size increases, new courses are added in many fields and ability grouping occurs in some fields; and in most of these courses, the cost per pupil is higher than the average cost of the particular field in which the course is found. The major reason for the higher costs seems attributable to a smaller average class size than is found in the existing courses for the subject field. Some of the new courses, which are frequently offered at an advanced level, tend to have smaller classes partly because educators use different input coefficients in their production functions. On the other hand, the primary reason for small average class size in most of the new courses seems to be that the demand for the new courses is limited, at least at the beginning of the operation of the course.

Services resulting from an increase in diversification within the supportive services, especially in the two large schools in the high net operating expense range appear to be more expensive than "regular" services. The two large schools provide not only a broader scope of supportive services, but also a greater variety of services. Both of these dimensions of diversification result in some role specialization, (for example, the employment of research directors), and the introduction of certain indivisibilities such as bookstores and workrooms. An increase in diversification within the supportive services in the schools in the low net operating expense range is not pronounced, and a consequence seems to be that the per pupil costs decrease slightly as size increases.

Indivisibilities

Even if it were desirable to do so, it would not be possible to increase inputs in exactly the same proportion as outputs are increased. This is because inputs come as discrete units, and not as infinitely divisible quantities.

Units of space provide one example of the indivisibility of inputs. Consider a one room school, built for 30 students. It is impossible to add a third of a classroom when the enrollment increases to 40. Rather, an entire new classroom must be added, with a total space for 60 students (if the 30-student size is maintained). In the case of a large secondary school, it is not feasible to add a single classroom. Rather, if an addition is built to the school, it must for the sake of practicality be a section consisting of a number of classrooms and supporting services such as light and heat. (One exception to this rule is the use of mobile classrooms, which permit adding one room at a time to even the largest schools.) Other space units, such as cafeterias, auditoriums and libraries, are, to an even greater extent than classrooms, all-or-nothing aspects of the total school plant; it is impossible, for example, to add to the library a few square feet at a time as the school's enrollment grows.

People, also, are indivisible. It is impossible to add continuously to the staff to keep up with growing enrollment. That is to say, we cannot add half an English teacher when the enrollment in English increases by only 15 students (unless the teacher is hired half time, or shared with another school; possibilities which are sometimes, but not always, feasible). Other personnel are even better examples of the phenomenon of indivisibility—for example, the principal, librarian, deans, and engineer.

Indivisibilities are integrally related to increasing returns to scale. As inputs, in the form of space, personnel, or equipment are added to a school, it pays to use them to the maximum. Hence, given a certain set of inputs, the cost per student hour of output decreases as these inputs are used more and more intensively. For example, a school may be planned to serve an ultimate school population of 2500 students. Although the first wing which is built will accommodate only, say, 1000 students, it is desirable to build certain facilities (for example, offices, library, and cafeteria) for the total student body, since these facilities cannot easily be expanded. As the student body increases in size, the cost per student of this space will decrease. Any school must have a principal; a school of 1000 students will no doubt need the services of a full-time principal. However, as the size of the school increases to 2500 students, it is not necessary to appoint a second principal. It may be that an initial half-time assistant principalship may change to a full time position; however, the increase in number of hours of administrative services required is proportionately less than the increase in student output-hours.

Specialization

It is a commonly observed phenomenon that as organizations grow in size, they are increasingly characterized by specialization and division of labor. Furthermore, division of labor often leads to additional efficiency in production, since each worker gains expertise in the performance of a specialized aspect of the total productive process.

Division of labor may be horizontal or vertical. Horizontal division of labor may mean that various aspects of the physical processes of production are performed by separate individuals. Vertical division of labor involves the separation of the processes of coordination and control from the processes of production.

Increasingly, division of labor may be expected to characterize large secondary schools. This division may take a number of directions. From the subject matter point of view, for example, an additional homemaking teacher may be employed when a school's enrollment increases from 1000 to 2000 students. The two teachers will then, presumably, specialize—the one teaching classes in sewing, and the other in cooking. This should lead to greater efficiency, since specialists can be employed, and each teacher can instruct in the subject area she knows best. Similar specialization can be anticipated in shop, science, business, and foreign language.

In the second place, there may be a division of labor according to the function a teacher performs, rather than the subject she teaches. For example, if the number of mathematics teachers increases from four to eight, one of these eight teachers may be appointed or elected team leader and may assume such tasks as supervision, coordination, and purchase of supplies. Some larger schools organize their teachers on a team basis, with a clear role distinction, and a separation of responsibilities. Most important, as the role of the teacher is rationalized, certain non-teaching functions can be assigned to clerical persons, teachers' aides, or theme readers.

Division of labor may also be manifested in the separation of decision making from teaching. For example, as the counselling function is better developed, counsellors may perform student advisement functions which were formerly performed by teachers. In a large organization, where teachers perform specialized roles, coordination becomes more difficult. This may call for the addition of more administrators, whose role also becomes specialized.

Specialization and the division of labor may result in an improved quality of education for the same or a lower cost. To the extent that this comes about, specialization leads to increased returns to scale. However, careful cost analysis is necessary to ensure that specialization, especially at the management level, does not merely result in a proliferation of roles, with no real relationship to organizational objectives.

Variety

Finally, increasing size is related to the variety of offerings. For example, consider the possible shift in foreign language offerings in school A when it changes from 1000 to 2000 enrollment (Table 3-8).

The resulting increase in variety does not necessarily change the per student cost. Hence, it may be possible for a larger school to provide a more varied output for the same total expenditure. This is another form of *increasing returns to scale*.

Table 3-8 Hypothetical Shift in Foreign Language Offerings (School A)

1000 Enrollment		2000 Enrollment	
French I	6 sections	French I	10 sections
II	4 sections	II	5 sections
III	2 sections	III	3 sections
IV	1 section	IV	2 sections
Spanish I	2 sections	Spanish I	4 sections
II	2 sections	II	3 sections
		III	2 sections
		IV	1 section
		German I	1 section
		II	1 section
		Latin I	1 section
		II	1 section
Total	17 sections	Total	34 sections

It should not be concluded that increasing size will, without limit, result in improved productivity, since other types of outputs must be taken into consideration. In particular, it is important that the advantages of personalized instruction, and of a sociopsychological climate that is favorable to interpersonal interaction, not be lost. However, certain types of organizational procedures (such as "schools within schools") have been devised to combat depersonalization in large secondary schools and universities.

Chapter Four

❖❖❖❖❖❖❖❖❖❖❖❖❖

Theoretical Approaches to Resource Allocation

This chapter uses the "psychologist's production function" for developing some theoretical approaches to resource allocation in educational systems. The chapter is more abstract than the remainder of this book. However, the tools which are described here are the implicit underpinnings of many aspects of administrative practice. It is suggested that administrators-in-training should become familiar with the process of resource allocation, as a basis for fiscal decision making. Such procedures as cost-effectiveness analysis and program budgeting are based on the concepts of this chapter. Furthermore, the analyses presented in the following pages provide a link between empirical input-output studies and administrative practice.

Our goal is to advance the development of a management science in education. The achievement of this goal is to be based on an analysis of educational inputs and outputs, and of the relationships among them. The analysis is based on certain assumptions, which will be explicated in this chapter. The fact that these assumptions are not empirically validated does not destroy their value. Their usefulness lies in the hypotheses we can draw from the theory which underlies the assumptions.

This is an important caveat, because our assumptions are difficult to validate; they are statements that are necessary in order that the analysis may proceed rather than well-verified propositions. However, the assumptions are not without foundation. In the first place, they provide the basis for much of the practice of educational administrators. In the second place, they are grounded in the literature, although not as specifically as they are stated below. In the third place, there is some empirical justification for them, although the evidence in some cases is mixed.

Instead of student hours of a given service, outputs are now defined in a manner analogous to the "value added" approach used by businesses and manufacturing concerns. In industry, value added is the increase in the value of a product at each stage of the manufacturing process. Value added is an appropriate output measure, because it takes into consideration both quantitative and qualitative factors. The value added at each stage in the manufacturing process can be related to the materials used (which are in turn outputs from the previous manufacturing process), as well as to labor, and the use of factory equipment, facilities, and overhead costs.

In education, value added is defined as the amount of additional learning which takes place at each stage of the process. If learning is defined as consisting of the mastery of discrete units of skills, knowledge, or understanding, value added is the number of these units acquired at each stage. If the learning is conceived as being continuous, involving the synthesis and integration of subject matter, value added is the difference between the level of understanding reached at a given stage and that reached at the next stage.

The term "value added" fits well into the context of human capital formation. The amount of educational capital which an individual possesses may be defined as the learning he has completed; hence, value added consists of increments in learning. At a given point in time, an individual can invest his capabilities in obtaining employment which provides a financial return, or he can invest them in additional education, and ultimately a higher level of income.

We can now discuss the concept of value added in terms of increments in educational performance. Such increments may be measured as follows. First, a sample of behavior in a given area of the curriculum is observed and measured. Second, by similar means and using the same subjects, a second sample of behavior from the same curricular area is measured after a lapse of time. The performance increment is the difference between the first and second measures.

In symbolic terms:

p = performance at time t
p' = performance at time $t + 1$
$p' - p$ = the performance increment in a single unit of time.

This procedure can be carried out for individual students. However, it is probably more useful for our purposes to use class or school averages as measures of p and p'. These averages can then be used as a basis for the development of decision models.

ASSUMPTIONS 1 AND 2 This analysis is based on the assumptions that (1) performance can be measured, and (2) the measurement procedures which have been developed by psychologists working in the field of education are adequate for the following procedures.

This assumption is, of course, essential in order that we may proceed further. It is based on the fact that authorities in the area of measurement in education have given a good deal of attention to the problems which are implicit in measuring performance and performance gains.[1]

By assuming that it is possible, by present techniques, to adequately measure performance, we have, as specialists in education will recognize, divested ourselves of some of the most serious difficulties in this analysis. We take a moment to point out some of the problems which exist in educational measurement, and which represent possible weaknesses in this assumption.

In the first place, educational systems have many objectives, and it is difficult (a) to measure performance along all dimensions of educational objectives, and (b) to provide suitable weightings to indicate the relative importance of the various objectives. Hence, we are faced with this question: What performances should be measured, and what weights should be given to the various aspects of organizational performance?[2]

---

[1] See for example, Benjamin S. Bloom, *Stability and Change in Human Characteristics* (New York: Wiley, 1964), pp. 113–115. Despite the impressive advances which have been made in the measurement of aptitude and performance, psychometrics remains an imprecise science. Even more difficult is the development of the scales which would make it possible to measure performance *increment,* which can be treated mathematically. One possible method is to convert raw scores to "grade equivalents."

[2] This problem is not unique to education. Although it is often assumed that businesses are run with the sole purpose of making a profit, businessmen have, in actuality, many objectives.[3] Churchman *et al.* suggest one procedure for specifying the objectives of a business firm and weighing them according to their relative importance.[4]

[3] Richard M. Cyert and James G. March, *A Behavioral Theory of the Firm* (Englewood Cliffs, N. J.: Prentice-Hall, 1964).

[4] C. West Churchman, Russell L. Ackoff, and E. Leonard Arnoff, *Introduction to Operations Research* (New York: Wiley, 1957), pp. 108–109.

Even after we have defined our objectives and ordered them in some system of priorities, there remain some problems in providing reliable and valid measures of performance. It is more difficult to measure certain behaviors than others. The affective and higher cognitive aspects of education are difficult to assess, while it is relatively easy to measure the acquisition of knowledge and the development of the lower order cognitive skills. Assessment procedures in education probably tend to overemphasize the measurement of those behaviors which can be readily observed and evaluated.

A more technical problem has to do with the development of measurement scales, which are useful in statistical analysis. The following discussion is based on the assumption that cardinal scales can be developed that represent performance in a given subject or skill area.

While we treat performance increments as outputs of educational systems, we must recognize that much learning takes place outside the confines of the school. Both desirable and undesirable behavior may be developed in the home or in other aspects of out-of-school life, and performance gains cannot be regarded as entirely the product of school effects.

All of these problems throw some doubt on the validity of the first two assumptions. However, we proceed in the hope that the analysis itself is sound, and leads to important implications for practice and for research.

THE PRODUCTION FUNCTION

Our first two assumptions were that the outcomes of education can be measured. In the following paragraphs, we present some additional assumptions. The total set of assumptions then forms the basis of the analysis.

ASSUMPTION 3 There exists a production function which relates outputs (performance increments) to inputs (students' time, teachers' time, equipment, including materials and books, and space).

Symbolically, we write:

$$p' - p = f(x_1, x_2, x_3, x_4, x_5)$$

p' and p are defined above
x_1 is student hours

x_2 is teacher hours
x_3 is equipment
x_4 is space
x_5 represents other, nonidentified variables

The term "function" indicates mathematical dependence. The statement, therefore, means that performance increments depend upon (or are related to) inputs of time of students, teachers and others, equipment, and space.

From a psychological point of view, the totality of "inputs" is the environment to which the child is exposed. Since there are often several educational environments in play at a given time, the production function must be thought of as being sufficiently complex to take account of a multiplicity of variables.

Educational practice is based on an implied production function, representing the effects of forces at work within schools or colleges. In order to bring about the changes which are considered to be educationally desirable, educators build schools, hire teachers, and purchase equipment, materials, and books. Administrators of secondary schools schedule teachers and students to occupy classroom space and to use equipment at certain times and in certain subject matter contexts. Attempts to improve performance usually involve increasing inputs by lengthening the school day, offering summer school programs, reducing the pupil-teacher ratio, purchasing films, or selecting and buying books. In addition, inputs of improved quality are sometimes substituted for inferior inputs in an attempt to improve outputs.

ASSUMPTION 4 All other things equal, increases in x_1 and/or x_2 are associated with increases in the increment $p' - p$.

It is assumed that if larger amounts of student time are devoted, say, to Algebra 1 (assuming no decrease in motivation or the other factors related to learning), the performance increase will be larger.

A more tenuous assumption is that larger amounts of teacher time are associated with increased learning. The evidence about the effect on learning of changes in class size is mixed. However, our assumption will be that at least in some areas—such as affective learning and the mastery of higher order mental processes—additional teacher time will result in improved learning, other things (including the quality of teaching) being equal.[5]

---

[5] The very concept of "class size," which appears to accept the notion of a fixed ratio of students to teachers, is probably antithetical to the development of more flexible production functions. The following discussion assumes that appro-

The other input variables are regarded as being of secondary importance. Inputs of equipment, books, and materials, x_3, will be considered as being primarily complementary to teachers' time, x_2. In other words, teacher effectiveness can be enhanced by the optimal use of material inputs. However, other inputs, such as television, teaching machines, and the like, will be treated in the analysis as possible substitutes, under some conditions, for teachers' time.

Finally, x_4 will not be regarded as contributing to performance. The lack of appropriate teaching and learning spaces may hinder effective instruction. However, space will not be considered as being an element which contributes directly to the learning process, adequate space being a necessary but not a sufficient condition for learning to occur.

A mathematical representation of an education production function can be developed in the following way. Select a sample of schools which vary along the input dimension. Through multivariate analysis, determine the relationship, for this sample, between input variations and output increments. Express this relationship, in probabilistic terms, as an input-output relationship. The functional relationships are likely to be nonlinear. That is, increases in a given input will not always lead to uniformly proportional increases in output. For example, the first thousand books in a school library may contribute more than the tenth thousand books. Hence, estimating input-output relationships will probably mean estimating the form of the function as well as its parameters.

Another method of arriving at an education production function would be to select a single school and to conduct a number of experiments consisting of changing input levels and observing the effect on outputs. If a sufficient number of experiments were carried out, a result similar to that described above would be obtained.

Production functions can be improved through a *search* for more efficient methods of instruction and organization. Some of the inputs

---

priate procedures are combined with variations in class size, and that under these conditions, learning takes place more readily when an intensive use is made of teacher inputs. On the other hand, it is possible that certain outputs, such as improved ability in the more mechanical aspects of arithmetic can be achieved more efficiently through the use of other inputs, such as computers or programmed texts.

Also, the following rather mechanical analysis does not deny the importance of sociopsychological processes. This chapter outlines a theory of resource allocation, which can then be used as a basis for considering one dimension of the problem of organizing educational systems. The effect of sociological and psychological variables on the economic model can then be examined.

would be spent on search rather than on the production of outputs. It is likely that the level of outcomes obtained in this way would be higher than that obtained if all the inputs were directly applied toward the production of outputs.

The search for knowledge need not consist entirely of school system research. The system might, for example, hire teachers and administrators whose previous training and experience have provided knowledge which is useful in the employing school. Alternatively, the system may devote some of its resources to the restraining of people who are presently employed. Finally, research personnel may be hired. The purpose of such people would be (a) to facilitate the use by the system of knowledge which has been developed eleswhere, and (b) to produce new knowledge through research within the system.

From a behavioral point of view, the study of educational production functions must include an examination of the manner by which aspirations and expectations are developed. Aspirations, at the organizational level, are defined as performance goals which are shared by members of the organization. Aspirations may result from past performance. They will also be affected by events outside the organization which result in a change in the performance level that members see as being desired or necessary. If performance reaches the aspiration levels (for example, if an English teacher believes that the work of her students reaches the standards that she sets), satisfaction ensues. If performance is below aspirations, there is dissatisfaction. The result of dissatisfaction may be that the organization will turn to search procedures to attempt to locate improved means of reaching a set of goals. Alternatively, it may solve the problem by lowering its aspiration level. Problems associated with aspiration level are present when the organization is affected by environmental changes. For example, if the school's clientele changes from middle to lower class as a result of population movements, the previous aspiration levels may be too high, in comparison with outcomes which are realizable in the short run. Also, when advances in knowledge in the larger society place higher demands on the goals, previously set aspiration levels may be too low.[6]

Expectations are sets of beliefs about the relationship between inputs and outputs. These beliefs may, in the exceptional case, have some statistical foundation. In most cases, however, they have been developed as a result of the past experience of members of the organization. Organizational learning consists, in part, of a revision of expectations among

[6] James G. March and Herbert A. Simon, *Organizations* (New York: Wiley, 1958).

members of the organization. Since the personnel in a school represent a number of coalitions, with somewhat different sets of expectations, the expectations of the total organization consist in large part of compromises among the expectations of these coalitions.[7]

The aspirations and expectations of members of the system will affect both the objectives which are chosen, and the degree of success the system has in meeting these objectives. These concepts can be considered to be caveats to the following analysis, which develops a theoretical approach to the input-output relationship.

MARGINAL ANALYSIS

We now proceed to apply some principles of microeconomic analysis to the study of the internal operation of schools. The essence of the analysis is the study of the effect of small changes in the independent variables (inputs) on the dependent variable (outputs). These changes are called marginal changes, and the analysis is referred to as marginal analysis.

The production function is intended to apply to a school (or at least a classroom) as a whole. All the above variables are to be treated, therefore, as averages, and not as the attributes of individuals. Furthermore, the following analysis will ignore the effect of scale.

In the first part of the analysis, we will concentrate on variables x_1 (teachers' time) and x_2 (students' time). There are good reasons for emphasizing these variables. In the first place, they are the most costly resources which the administrator has at his control. Teachers' salaries typically comprise between 65 and 70 percent of the direct costs which are included in the school budget. Furthermore, this is the cost item which is increasing most rapidly. With respect to students' time, Schultz and others have pointed out that this input comprises over half of the total direct and indirect cost of education in secondary schools.[8] From the educational point of view, these are the variables which contribute most to learning.

The allocation of the time of students and teachers is therefore the most important single task of the school principal. The process is as follows. First, the curriculum is broken up, usually according to subject

[7] *Ibid.*, ch. 4.
[8] Theodore W. Schultz, "Education and Economic Growth" in *Social Forces Influencing American Education* (Chicago: University of Chicago Press, 1961).

areas. Second, within each area, the time of specialist teachers and supporting personnel and the time of students are allocated. The allocation of space and equipment is coordinated with that of personnel. Usually, in the past, scheduling has been a once-a-year task. Today, with increased use of computers, and with the recognition of the importance of the scheduling process in adjusting means to ends, monthly or weekly scheduling seems both possible and desirable.

In Figure 4-1, the point C represents a weekly allocation of 5 hours of students' time and 0.25 hours of teachers' time. For 20 students, this would involve a total of 100 hours of students' time and 5 hours of teachers' time.

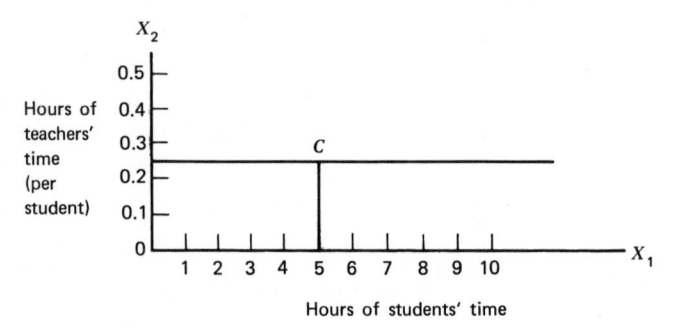

Figure 4-1 Allocation of time for Biology I.

Subject matter areas (or university departments) compete for resources. The Chairman of the English Department, for example, will argue that an additional English teacher is more important for the high school than an additional teacher of physics or homemaking. Equally important, subject areas compete for the time of students. This competition is manifested in disagreements over the number of courses to be offered in each subject area, and the number of minutes per day and weeks per year to be allocated to each course.

One way to obtain additional student time is through homework assignments. Teachers in the various departments of a secondary school will compete with each other for students' out-of-class time. An uneasy compromise, which provides for sharing students' time among the various subjects, may be reached. For example, each subject area may be allocated a number of hours of students' home time per week, or certain days of the week may be set aside for homework assignments in each subject.

Teachers also spend time in addition to the time spent in class. In the first place, most teachers need time to prepare lessons, grade papers, and assist students, so that an hour scheduled with a class involves more than an hour of total time. In the second place, many school systems schedule some teachers' time to the improvement of their own knowledge and skills.

We will now turn to the relationship between input mixes and the output of an educational system. Consider first the assumption that, other things being equal, an increase in teachers' time (per student), or an increase in students' time, or an increase in both, will lead to an increase in output.

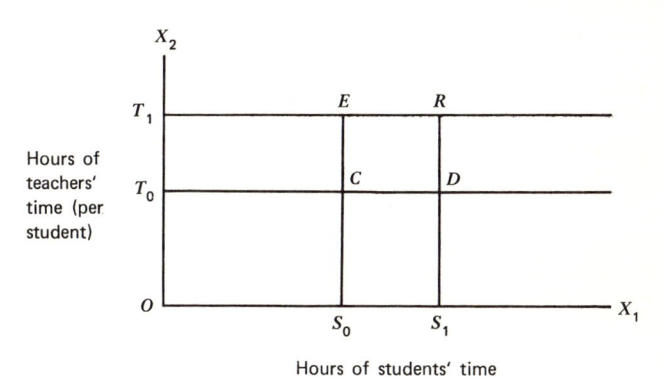

Figure 4-2 Changes in input variables.

In Figure 4-2 an increase in students' time from OS_0 to OS_1, given an input of OT_0 units of teachers' time, is reflected by a change in the input mix from C to D. A change in teachers' time from OT_0 to OT_1 given an input of OS_0 units of students' time results in change in the input mix from C to E. Finally, if the input mix changes from C to R, students' time increases from OS_0 to OS_1 and teachers' time increases from OT_0 to OT_1. Each of these changes would, according to the assumption, be accompanied by an output increment.

The move from C to D, representing an increase in students' time while teachers' time remains constant, could represent an increase in the out-of-class work which is done by students. If students spend the same amount of time in class and spend more time in the library or in doing home assignments, it may be expected that, on the average, and assuming no decrease in interest or motivation, performance will improve.

The move from C to E, on the other hand, might represent a reduction of class size with no change in the time students spend on their work. Alternatively, it could represent more use of teachers with some small groups and individual students, without a change in the size of the instructional unit. Our assumption is that, over broad ranges of the teacher time variable, increases in teacher inputs per student hour, with other variables (including the quality of teachers) held constant, will improve performance.

Finally, the move from C to R represents increases in both inputs. Assumption 3 would lead to the conclusion that output is greater at R than at $C, E,$ or D.

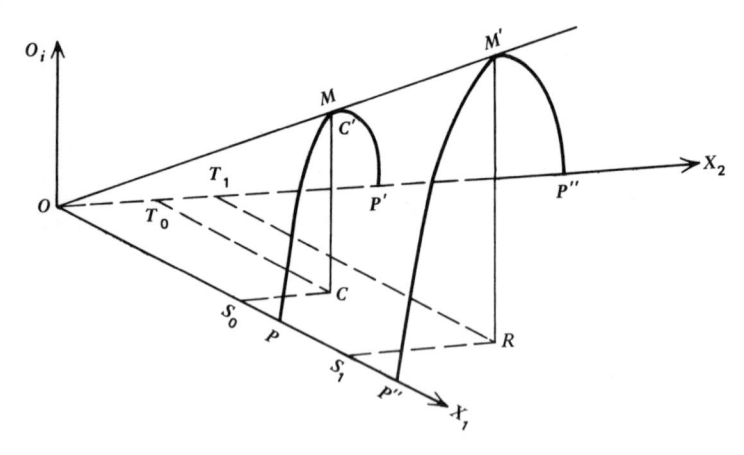

Figure 4-3   Relationship between output O_i and mixes of inputs X_1 and X_2.

We now turn, in Figure 4-3 to an analysis of the relationship between output levels for output O_1 and mixes of inputs X_1 and X_2.

In Figure 4-3 we have turned Figure 4-2 on its side, so that units of variables X_1 and X_2 are shown on the horizontal plane, while units of the output O_i are shown along the vertical axis. Take the point R, as in Figure 4-2, and erect at R a line RM', whose length represents the amount of output which is associated with the inputs (OT_1 and OS_1) which R represents. If the same is done for every point representing a combination of X_1 and X_2 the result is a curved surface, whose height above the horizontal plane represents the output associated with each input combination.[9] OMM' (which need not be a straight line) joins points of maximum output (associated with increases in inputs).

The shape of the "production surface" illustrates our two assumptions. In the first place, an increase in inputs (say, from C to R) results in an increase in outputs. In the second place, there are some input combinations which are more effective than others. At the extreme, a point on the line OX_2 represents teachers' time in conjunction with zero units of students' time—a combination which may be assumed to lead to zero units of increase in student performance. On the other hand, a point on the line OX_1, representing students studying without the help of teachers, cannot be assumed to lead to zero units of performance increments. However, we will assume that some combination of teachers' and students' time will produce an optimal rate of student progress.

The diagram in Figure 4-3 is much oversimplified. The cross section PMP' need not be symmetrical, for the two inputs need not make similar contributions to output. The cross section PMP' need not be smooth—it may have dents, kinks, or even sharp breaks.

Point M, the highest point in the cross section PMP' will not normally represent an optimal arrangement. It does represent a productive combination of inputs. However, it may not pay to use inputs in this combination. It must be remembered that teachers' time is expensive, and that resources must be used for obtaining other inputs. Furthermore, there is a limit to the amount of students' time which is available. Hence, we may have to be satisfied with a lower level of output than M.

One principle governing the level of output which may be obtainable with limited resources is the "law" of diminishing returns. It may be defined as follows. As more and more of a given input is employed, *all other inputs remaining constant,* the point is reached at which additional units of the given input will make decreasing marginal contributions to the output.

Consider the following examples:

1. TEACHERS' TIME

Suppose we have 1000 students registered for Biology I. If one teacher is hired, he will be very much overworked, and his contribution to students' learning in Biology I will be limited. A second teacher (quality of teacher held constant) will contribute more, a third teacher more, and so on. However, each teacher after the first adds less to total output

[9] See William J. Baumol, *Economic Theory and Operations Analyses* (Englewood Cliff, N. J.: Prentice-Hall, 1961), ch. 9.

than his predecessor and the point will finally be reached at which the contribution made by an additional teacher to performance in Biology I will be negligible. This is the phenomenon of decreasing marginal returns. (See Figure 4-4.)

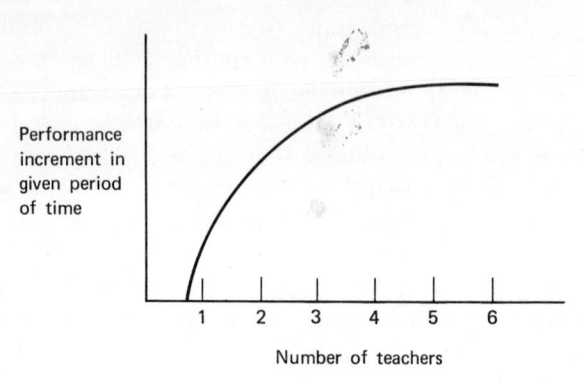

Figure 4-4 Number of teachers and performance in Biology I (for 1000 students).

2. STUDENTS' TIME

The assumption here is that, as students spend more and more time on a given budget of work, the added performance which results from increased time allocations eventually begins to decline. This statement is made with some hesitation, because it represents an intrusion into the realm of the psychologist. The type of learning which is involved is important, since areas which require an extensive synthesis of knowledge may require more time before the desired result is obtained. Furthermore, the statement does not, in its simplest form, take account of thresholds, below which learning is not appreciable. However, the assumption is necessary for this analysis. Its factual content can be appraised; inaccuracies in the assumption may affect the following sections of the analysis.

The relationship between students' time and learning may take several forms. (a) There is probably an optimal length of a lesson in a given subject for students with certain characteristics. (b) The number of minutes per week assigned to a specific subject is presumably related to the amount of learning which takes place. (c) The number of weeks a subject is studied in a given year, and the total years devoted to a subject are also important. Not enough attention has been paid to studying this aspect of the time allocation question. For example, more must be known about the relative merits of a short, intensive study of a foreign language versus the traditional pattern of five periods a week for several years.

Let us consider (b) above in greater depth. The time allocation problem is, in part, a problem in marginal analysis. Up to a given point, as more time is spent on a given subject, the amount learned per time unit may be assumed to increase. Beyond the point, each unit of time yields a successively smaller increment of learning (Fig. 4-5).

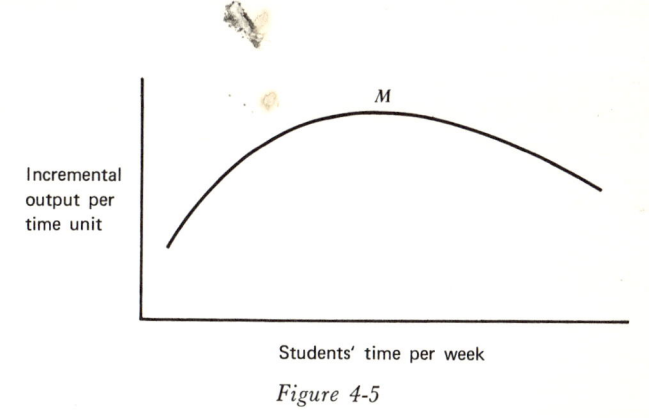

Figure 4-5

Up to time point M, each additional unit of time yields an increased increment of learning. An overcrowded schedule in which students do not reach the point of optimal marginal returns to their investment will lead to the inefficient production of education. Skillful scheduling and some individualization of time allocation are needed, so that students remain with a subject at least as long as the marginal product associated with their use of time is increasing.

3. EQUIPMENT

It is plausible that the law of diminishing returns applies also to physical inputs. Consider the addition of an overhead projector to promote effective learning for 1000 Biology I students. The improvement which results may be substantial. If a second overhead projector is added, further improvements may result. However, it would be foolish to continue indefinitely in the purchase of this kind of equipment, since a point will be reached at which the machines are idle a good deal of the time, and when an additional unit will merely take up more storage space, thus adding to costs but not to benefits.

The purchase of library books is a more ambiguous example. We may assume that library books complement the work of teachers, and that additional books permit students to go far beyond the knowledge teachers possess. However, a point would eventually be reached when so much is spent on library books that other important inputs are in short supply. It is not feasible to consider inputs in isolation. Rather, we must consider how a given sum of money should be allocated among a number of inputs, considering that the law of diminishing returns applies to each. To simplify the analysis, we consider the problem of allocating resources between only two inputs. In the following discussion we assume that technically efficient input combinations have been identified.

Figure 4-6 is called an isoproduct curve. This is a line which joins points representing input combinations which make equal contributions to output. We consider combinations of teachers' time and books, as they contribute to performance in English literature. In this hypothetical example, line LL' suggests that it is possible to substitute books for teachers (or teachers for books), so that the same outcome is obtainable with, say, 5000 books and 800 hours of teachers' time as with 1000 books and 3000 hours of teachers' time.

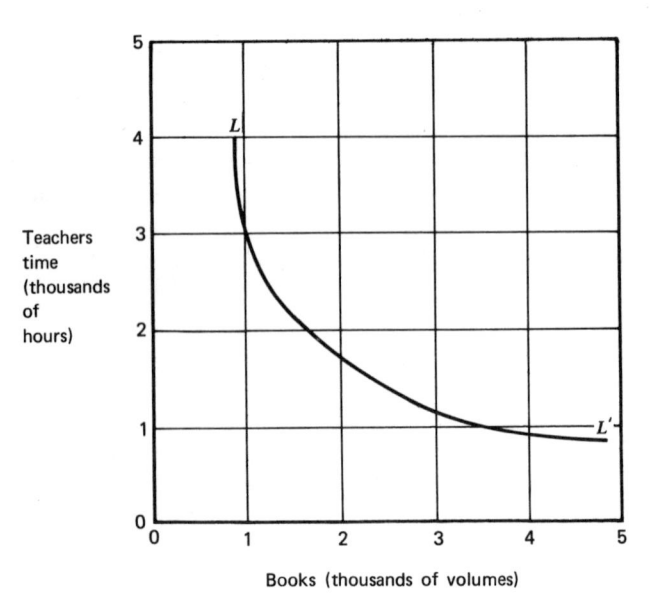

Figure 4-6 Production isoproduct curves. Weekly inputs of teachers' time and books (for 1000 students).

Other curves represent other performance levels. One can choose the curve which represents the performance level which is aspired to, and examine the various combinations of inputs which that curve represents. The shape of the curve (convex to the origin) suggests that at *higher* levels of one input, it requires less of the second input to substitute for it. For example, the following figures show the amounts of books associated with different amounts of teachers' time in production function LL'.

Teachers' Time	Books
1000 hours	3.4 thousand
2000 hours	1.6 thousand
3000 hours	1.0 thousand
4000 hours	0.9 thousand

A reduction in teachers' time from 4000 to 3000 hours requires an increase of only 0.1 thousand books, while reduction from 2000 to 1000 hours requires an increase of 1.8 thousand books. In terms of books, the first hours of teachers' time are more valuable. This is, of course, an application of the *law of diminishing returns*.

Economists have made extensive use of curves of this type (called isoproduct curves). By deductive reasoning and indirect empirical tests they argue that normally such curves will have the following properties:

1. They have a negative slope. (An increase in input of books can substitute to some degree at least for teachers' time.)
2. If one curve, X, lies above and to the right of another iso-product curve Y, X will normally correspond to a higher output level than Y. (More of *both* inputs will increase outputs.)
3. No two isoproduct curves intersect. (Output cannot be increased by reducing one of the inputs.)
4. The curves are convex to the origin.[10]

Given any level of one input, X_i, the increase in output which can be obtained by a small increase in X_i (with other inputs held constant) is called the marginal product of X_i. Note that, due to the law of diminishing returns, the marginal product of X_i will decrease as X_i increases. One mathematical property of the production iso-product curve is that its slope, at any point, is equal to the ratio of the marginal product of the two inputs represented on the curve. This ratio is a measure of the rate at which one input may be substituted for the other, while the output remains constant.

[10] Baumol, *op. cit.*, p. 183.

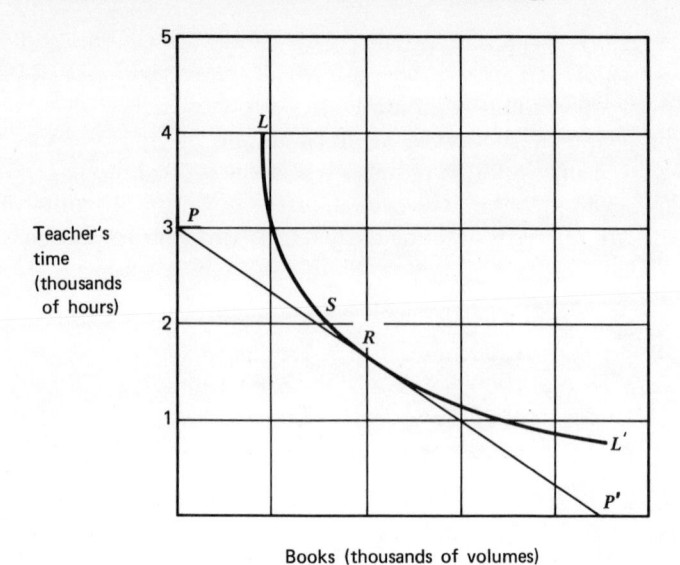

72 THEORETICAL APPROACHES TO RESOURCE ALLOCATION

Figure 4-7 Production isoproduct curve and price line.

The above analysis has omitted discussion of the fact that some inputs are more expensive than others. To make the story complete, we redraw, in Figure 4-7, the production iso-product curve, with the addition of a price line. To obtain the latter, we assume that books cost, on the average, $5.00 per volume, and that teachers' time costs $7.50 per hour. Assume that $22,500 is available for books and teachers. At the above price, this will purchase either 4500 books or 3000 hours of teachers' time, or any combination of the two which sums to the amount of money available. The line *PP'* in Figure 4-7 shows this combination of possibilities.

The point of optimality (for given expenditure level) is *R*, where the price line is tangent to the production isoproduct curve. At *R*, the slope of the price line is identical with that of the curve. Since the slope of the curve is equal to the ratio of the marginal products of the two inputs, and the slope of the line is equal to the ratio of prices of one unit of each input, we obtain the following basic optimality rule:

An optimal combination of any two inputs, *U* and *V*, requires that the ratio of their marginal products be equal to the ratio of their prices. Symbolically,

$$\frac{MP_u}{MP_v} = \frac{P_u}{P_v}$$

The equation may be rewritten: $MP_u/P_u = MP_v/P_v$. Suppose, for example, that one additional unit of input u (teachers' time) produces a performance increment of 15 points, and costs $7.50, while an additional unit of input v (books) produces a performance increment of 12 points, and costs $5.00. One dollar spent on input u therefore brings a return of 2 points increase in performance, while a dollar spent on input v brings a return of 2.4 points. It pays, therefore, to shift some expenditures from input u to input v, which is the more productive at this point on the production curve.

This analysis can be extended to the multiple-input case. *Resources should be allocated among the various inputs in a school, until the marginal product divided by the unit price is the same for each input.* If it were possible to apply this rule, it would then be a relatively simple matter to determine how much should be spent for each of the inputs—books, equipment, teachers' salaries, and so forth—that contribute to students' learning.

One problem is that it is generally impossible to *know* how much will be added to output by the addition of a unit of given input. The best we can usually do is develop what are known as *subjective probabilities*. These are personal judgments, based on past experience and on whatever data are available. These probabilities can be refined, as additional data are obtained.[11]

In Table 4-1 we illustrate one procedure which may be used. Note

[11] For discussion of methodology see Donald L. Meyer, "Bayesian Statistics," in *Review of Educational Research*, **XXXVI**, No. 1405 (December, 1966), ch. II. Also, Herman Chernoff and Lincoln E. Moses, *Elementary Decision Theory* (New York: Wiley, 1959). R. Duncan Luce and Howard Raiffa, *Games and Decisions* (New York: Wiley, 1958).

Table 4-1   Illustration of Procedure for Resource Allocation

Input Variable	Cost per Unit	Less than 0.7	0.7–0.89	0.9–1.09	1.1–1.29	More than 1.3
A	$5	0.1	0.2	0.4	0.2	0.1
B	2	0.3	0.4	0.2	0.1	0
C	6	0.1	0.1	0.2	0.4	0.2
	Midpoint of Interval	0.6	0.8	1.0	1.2	1.4

Expected Performance Increment per Unit (Subjective Probabilities)

that by using subjective probabilities, the analysis becomes possible even though the ratio of marginal product to price is unknown. The procedure in fact probably formalizes what is done implicitly by the rational administrator.

We can now calculate the performance increment associated with the addition of a unit of A, B, or C.

A: $(0.1 \times 0.6) + (0.2 \times 0.8) + (0.4 \times 1.0) + (0.2 \times 1.2) + (0.1 \times 1.4) = 1.0$

B: $(0.3 \times 0.6) + (0.4 \times 0.8) + (0.2 \times 1.0) + (0.1 \times 1.2) + (0 \times 1.4) = 0.82$

C: $(0.1 \times 0.6) + (0.1 \times 0.8) + (0.2 \times 1.0) + (0.4 \times 1.2) + (0.2 \times 1.4) = 1.10$

We now divide the *expected marginal product* by the price, with the following results:

A: $1.0/5 = 0.2$
B: $0.82/2 = 0.41$
C: $1.1/6 = 0.18$

In this example, B is the preferred input, and it will pay to shift resources from A and C to B until the marginal products are approximately equal.

In considering the feasibility of applying this procedure, it must be remembered that we are not, of necessity, attempting to rationalize the entire allocation system. Rather, we should consider possibilities for making minor shifts in allocation in a given year—spending a few hundred dollars more for science equipment or for library books, hiring a specialist to teach reading, or offering a summer course for advanced students in the physical sciences. In each of these possibilities, an estimate may be made of the marginal product, and this estimate may be compared with the cost figures. If the effects of the program are carefully assessed, better estimates should be possible the next time similar plans are implemented.

The above analysis shows that it is advantageous to shift resources from the more costly to the less costly inputs. In industry, such shifts have often involved spending more money for equipment and machinery and less for personnel. In government services, and especially education, such shifts are not easily made. Education, and particularly the more human aspects of education—the development of affective outcomes, and teaching for competence in the higher order mental processes, depends upon the contributions made by people. The personal inputs into education are becoming more expensive, since: (a) higher qualifications are being demanded of teachers; (b) at present salary levels, shortages of certain categories of teachers persist; and (c) due to the increased strength of their unions and associations, teachers have been able to exert greater pressures at the bargaining table. Increased productivity in education depends on a continued awareness of the high cost of personal inputs—the time of teachers and students. Professor Theodore W. Schultz has cogently expressed this problem.

> The total value of all the time of the faculty and other university inputs entering into undergraduate instruction is not as large as the value of the time the student is putting in. The sacrifice, the opportunity costs of the students' time exceeds by a substantial margin all the rest. And yet we treat the time of the student as if it were of zero value, a free input. He should be forever grateful for being allowed to be around even though he is giving up three to four thousand dollars a year! This underevaluation of the student's time is a very serious matter, and there will have to come a reckoning. It has become more serious as our society has become richer because of the increases in earning power of persons who are qualified to attend college. This misallocation of the time of students seems to be getting worse instead of better. The reason is that the student's time is becoming more valuable in relation to some of the other components of education.
>
> The value of the time of the faculty is next, and here, too, we have not really faced up to the fact that the services of the faculty have become more expensive relative to other instructional inputs. We have not really looked for substitutions. As economists, we tell people to be alert to substitution possibilities among inputs as price ratios change. We tell them to use the cheaper inputs a little more in order to economize on the more expensive. Faculty time has become expensive relative to other teacher inputs, such as the library service. But the faculty spends an inordinate time in classrooms.[12]

The above discussion describes a theoretical basis for allocating resources, in order to obtain a given set of educational objectives. Another, and equally difficult problem remains. This involves analyzing the manner in which resources should be allocated among a number of sets of purposes which are not always compatible. We may ask this question in a number of ways. For example, instead of questioning how resources should be allocated to produce proficiency in mathematics, we can at-

---

[12] Theodore W. Schultz, "Teaching and Learning in Colleges of Agriculture," *Journal of Farm Economics*, **47**, No. 1 (February, 1965) p. 20.

tempt to determine how resources should be allocated, for example, among several subjects, such as English, foreign languages, mathematics, and industrial arts. We can also discuss questions of allocating resources among the brilliant, average, and slow students. Finally, states are increasingly having to defend financial procedures which discriminate in favor of high income communities.

These problems can be discussed in a number of ways. Since they are strongly normative, they may be subjected to philosophical analysis. Questions such as "What is the nature of man?" and "What is the nature of society?" affect the types of curricula we will use, and the way in which we attempt to educate various groups in the society.

Philosophical analysis will not, however, provide us with a single answer to the questions we are asking. The humanist may decide that students should study Greek, Latin, and ancient history, while the pragmatist may opt for modern languages, mathematics, and science. The philosopher who has a predisposition to elitism will emphasize the education of gifted children, while the egalitarian may object to special treatment for any one group of students.

In short, an analysis of the philosophical issues involved in establishing educational objectives will result in the identification of various value positions, and their implications. However, such analyses will not solve the practical problem of determining how to allocate resources among the various curricula and groups of students, since values differ, and cannot be ranked in a manner which will satisfy all persons.

It would appear, rather, that objectives are determined on a more pragmatic basis. Interest groups are formed, both within and outside the educational enterprise. Each of these groups engages in political activities designed to serve the purposes of its members. For example, lay associations for the promotion of the education of handicapped children have been successful in persuading legislatures to allocate resources for special education. Within the system, teachers of one subject (for example, English) compete with teachers of other subjects for good students, spacious classrooms, and a share of the budget for books and supplies. The resulting allocation constitutes a political accommodation quite in keeping with the manner in which other important decisions are made in our society.

Neither philosophy nor politics is quite compatible with the issues raised in this book. We are more concerned with economic optimality than with normative issues or the political process. According to our analysis, objectives will be selected which contribute most to the well-being of society and the individual. "Well-being" does not necessarily mean income; however, income is a useful variable to represent all the desired outcomes associated with education. We therefore apply the following rule:

> Allocate the available resources among alternate outputs, so that the marginal product (in terms of additional income for society) per dollar spent is the same for each type of output.

Consider the following examples.

1. Categories of students. According to this rule, money would be spent for the education of the gifted, the handicapped, and others, so that the marginal dollar spent for each group would bring the same returns in terms of additional income. While the rule is clear, its implications are not, since we have insufficient empirical evidence to permit us to decide to reallocate funds, for example, to gifted or to low income children, on the basis of this criterion.

In actual practice, the maximization of total income is not the criterion which would normally be used. For example, it may be desired to spend millions of dollars on the education of visually handicapped children for reasons which are humanitarian rather than economic.

2. Subject matter areas. Again, the rule is that money should be allocated among the subject areas of the curriculum in such a way that the marginal dollar spent, for example, on English, would bring the same additional income as a marginal dollar spent on foreign language or technical training. While we are again handicapped by lack of data, it would be possible to obtain data which would enable us to apply this rule to certain kinds of situations.

Suppose, for example, that we have evidence (from GRE scores) about the contribution which competence in English and mathematics will make to the total skills of school administrators. Again hypothetically, we find that superintendents who are very articulate do as well, because of their persuasive ability as quantitatively oriented superintendents who are in demand because of their ability to handle the financial aspects of the school situation. On the basis of these data, we are able to draw an iso-product curve which shows the contribution made by ability in English and mathematics to administrative competence.

We find also that in graduate school it costs $5 to add a unit of ability in English, but because of the expensive equipment and faculty time involved, it costs $7.50 to add a unit of ability in Mathematics. These prices constitute the basis of the price line in Figure 4-8.

In Figure 4-8, *OO'* is the output isoproduct curve, which represents combinations of educational outputs which make equal contributions to administrative success. For example, 50 units of English and 4 units of Mathematics (Point *A*) are equivalent to 20 units of English and

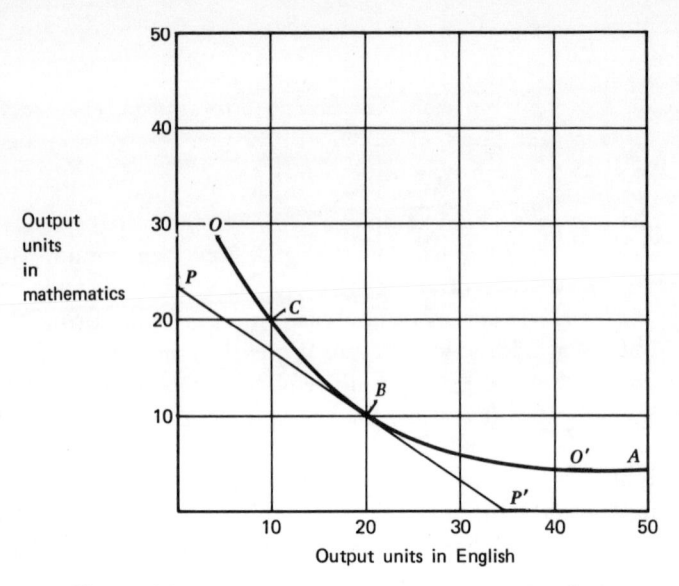

Figure 4-8 Output isoproduct curve and price line.

10 of Mathematics (Point B) or 10 units of English and 20 of Mathematics (Point C). Since the price line cuts the output isoproduct line at B, the optimal production, from the point of view of the graduate school, is 20 units of English and 10 of Mathematics. Note that this analysis is identical to the previous input analysis. This is not too surprising, since the "outputs" shown in Figure 4-7 are actually inputs which contribute to a given level of proficiency in earning an income.

This theoretical discussion leads to the applications now presented in the last two chapters. An understanding of the methods described above is particularly necessary, in order that the educational administrator understand the processes of resource allocation in which he is constantly engaged.

Chapter Five

Approaches to Decision Making

Textbooks in educational administration stress the importance of decision making; an effective administrator is able to define a problem, collect relevant data, and bring this data to bear on the problem. He identifies possible alternatives and selects the most promising. Finally, he evaluates the result of his decision, and is willing to modify it if the data so indicate.

Furthermore, the competent decision maker does not operate alone. He brings to a given decision all the expertise he can muster. Staff members whose knowledge is related to the problem being studied are involved in the decision process—especially if the decision is a major one.

One of the most promising signs in this literature is the new emphasis on planning. While practice at all three levels of government has traditionally been based on the one-year budget, it is increasingly recognized that important decisions require committing resources over a number of years. Program budgets, which require planning for a period of several years, are being implemented in many school districts. Information systems upon which long term planning depends are also being developed in some state and local school systems.

As the emphasis is placed on data-based decision making, there will be an increased need for formalizing the decision process. Objectives need to be identified, and solutions are required which promise the achievement of these objectives, if the time and energy which goes into careful planning is to be justified. This means that formal decision-making procedures must in some cases replace the more intuitive procedures which have prevailed in the past.

This chapter describes some decision-making models which have promise for educational administration. It begins with a discussion of input-output, cost-effectiveness, and cost-benefit models for decision making. It then discusses the use of probability models and feedback models in decision making. It closes with a discussion of information systems.

INPUT-OUTPUT, COST-EFFECTIVENESS, AND COST-BENEFIT MODELS

Each of these concepts implies a relationship; considerable research is being conducted toward providing an empirical basis for these relationships. However, this work has not yet reached the point where it provides a definitive basis for decision making. The problem is partly one of conceptualization—it is necessary to clarify the meaning of the terms and the nature of the mathematical expressions which underlie the relationships. It is also necessary to conduct research into the empirical underpinnings of the models, and to devise ways of using these results for improving decisions. The following section examines these models, with a view to enhancing their usefulness as decision-making instruments.

1. Input-Output Models

One important aspect of administrative decision making is resource allocation. The rational administrator allocates the resources under his control (including his own time and the time of teachers and students) so that the marginal products are equal for all inputs. The administrator needs some guides as to the relative value of each input; this is provided in part by the results of input-output analysis.

The appropriate studies consist of the multivariate analysis of cross-sectional data. A relatively large sample of schools is chosen; input and output data are collected for each school. Multiple regression equations indicate the efficacy of each input variable in the presence of the other input variables.

Research findings are not yet adequate to provide the basis for the decision models for which we are seeking. Additional studies should be conducted, with the goal of refining the types of production functions reported in Chapter 2. Additional empirical studies should be used in attempts to determine the degree to which input coefficients are stable when the composition of the sample is varied. The production functions implied by the regression equation should be disaggregated to report differences in regression coefficients among educational systems of varying size and with clients from differing cultural backgrounds. A master regression equation that is based on a total sample of all schools in the country would probably be of limited value as a guide to decision making, since variations within the sample call for different pedagogic and organizational techniques. Equations based on longitudinal studies instead of cross-sectional studies may be useful, especially when decisions are to be made regarding resource allocation within a given system.

The regression equation does not contain cost information. Even the teachers' salary variable, which is the closest approach to an independent variable that is also a cost factor, is a proxy for qualitative differences that are only partially due to salary. Hence, decisions based on the model should include considerations of the costs of using one combination of inputs (suggested by input-output analysis), as opposed to the cost of other combinations.

To summarize, decision procedures based on input-output studies should include the following steps.

(a) Calculate regression equations based on measures of inputs and outputs in a population of school systems similar to those within which the decisions are to be made.

(b) Study regression coefficients, and note the treatment effects which are subject to the control of policy makers.

(c) Examine the costs of alternative combinations of inputs and relate these costs to the predicted effect.

(d) Implement the alternative which has the most favorable predicted output-input relationship. Observe the results of this implementation.

It should be emphasized that the regression equations are based on probabilities. This should be taken into consideration in predicting the effects of a given treatment, since if the probability that a given treatment will produce the desired effect is p, this treatment will be ineffective on the basis of the prediction equation in $1-p$ percent of the cases. This probability of nonsuccess is over and above the error variance contained in any regression equation.

2. Cost-Effectiveness Analysis

Cost-effectiveness analysis is suitable for problems where the outputs of the system are not priced at the market, while the inputs are subject to market pricing.[1] Clearly, many decisions in education are of this nature. Suppose, for example, that a decision is to be made about the best procedure for teaching computer programming. Several alternatives are available: classroom instruction, classroom instruction supplemented by texts, and computer-assisted instruction. Each procedure has costs; these are valued at the market, in terms of the value of teachers' time, books, and time with the computer. Each procedure also has benefits; however, these benefits (the student's ability to program the computer for various types of problems) can be valued by the market only in the long run. It would therefore be appropriate, in making a decision, to measure the cost of each instructional system, and to measure the effectiveness of each, in terms of the student's capacity to write computer programs.

The procedures included in cost-effectiveness analysis consist of following five elements:[2]

(a) The objective. Cost effectiveness analysis is designed to facilitate the attainment of certain goals or objectives. Since most educational programs have more than one objective, it is helpful if agreement can be reached on a single objective, or a weighted index based on several objectives. Effectiveness is the measure of the extent to which the objective (or weighted index) is achieved.

(b) The alternatives. These are the various instructional or other procedures which are considered. One of the most important

---

[1] Thomas A. Goldman, editor, *Cost-Effectiveness Analysis* (New York: Praeger, 1967), p. 18. The distinction between cost-benefit and cost-effectiveness analysis is essentially that for cost-benefit analysis there must be some way of translating benefits as well as costs into monetary terms. To the degree that the result of an effective program is increased income, the distinction between the two forms of analysis becomes blurred. Mary Jean Bowman suggests that there are three essential analyses: (1) dollar cost by dollar benefit: simple cost-benefit analysis; (2) dollar cost for a fixed benefit (however, benefit is measured) or variable benefit for a fixed dollar cost: simple cost-effectiveness analysis; (3) dollar cost for a benefit that is realized in varying degrees, and has no measurement in dollars: either complex cost-benefit analysis or complex cost-effectiveness analysis. (In private memorandum to the author.)

[2] *Ibid.*, pp. 4–5.

steps in a cost-effectiveness analysis is to identify the relevant alternatives.
(c) The costs. The costs associated with alternative instructional systems (such as team teaching, the use of television, or computer assisted instruction) can be estimated. If the analysis is to be implemented over a period of several years, input prices may change, and the original cost estimates will need to be corrected.
(d) A model. The purpose of the model is to estimate the costs associated with each alternative, and the degree to which each alternative may be effective in permitting the chosen objective to be reached.
(e) A decision rule. The decision rule is a set of procedures which permits a choice to be made among alternatives on the basis of the cost and effectiveness of each.

There are few practical examples of the application of cost-effectiveness analysis to educational decision making. We turn, therefore, to some hypothetical examples.

EXAMPLE 1

A new high school is being planned for a suburban school district. The social studies department is studying the problem of selecting the best instructional procedures. In the first phase of their work, the members of the department list their objectives and develop methods for evaluating the accomplishment of these objectives. They define the objective on the basis of several factors, including achievement, the acquisition of desired values and the development of tools and habits associated with self-instruction. Three levels of the objective (high, medium, and minimum) are defined.

Three instructional methods are also identified. The first consists of group instruction, supplemented by books and traditional audio-visual aids. The second method supplements classroom instruction with extensive use of closed circuit television. The third procedure consists of reconstructing the role of teacher, introducing team methods of instruction, employing semiprofessionals to perform certain functions, and using closed circuit television to enhance the effectiveness of instruction.

Cost estimates are obtained for each type of instruction. Using the subjective estimates of teachers, supervisors, and administrators, these cost estimates are fitted to instructional procedures for each level of output which is considered. (While initial estimates are little more than educated guesses, these estimates can be revised, through an iteration process, as the program is implemented.)

Table 5-1 Cost for Attaining Each of Three Effectiveness Levels for Three Previously Defined Instructional Procedures

Instructional Procedure	Effectiveness Level		
	Minimal	Medium	High
1	$50	$75	Not possible
2	74	82	$97
3	87	91	95

Note: The instructional procedures are as follows:
  1. Group instruction
  2. Group instruction plus closed circuit television
  3. Team teaching, varied personnel input mix, closed circuit television

(There is, of course, no guarantee that the order of effectiveness of the three procedures is as shown above. In fact, the procedures would probably be differentially effective in different school systems, and with students from different backgrounds.)

Either Table 5-1 or Figure 5-1 may be used in the cost-effectiveness analysis. This analysis proceeds as follows:

1. Assume that the objective is to reach effectiveness level described as "minimal." This may consist of ensuring minimum adequate performance in the basis skill performances. Concomitant attitudes and habits of independent study are included in the determination of the index.

   It is seen that all three instructional procedures can be used, but that the first method, which consists of traditional types of instruction, is least costly. It should, therefore, be chosen.

2. The "medium" level of effectiveness may include sufficiently high achievement scores to permit students to develop additional competences such as ability in the interpretation of maps and charts, writing and reading skills, and leadership capabilities. It may also include the objective of preparing a large proportion of the student body for success in college.

   Higher input levels are assumed to be needed, regardless of the instructional procedure which is chosen, in order that this level of effectiveness may be reached. However, alternative one is still preferable.

3. The third alternative may include the development of the kinds of competences which are required for success in the prestige colleges. (The operational definition of this objective is left to the reader. In our illustration, alternative 3 is not attainable using the first instructional procedure, and the third alternative is preferred, since it is less costly than alternative 2.

This analysis is completely hypothetical. We are aware, in the first place, that students' backgrounds will help determine how easy or difficult it is to develop given levels of competence. There are so many interactions between students' backgrounds and instructional procedures

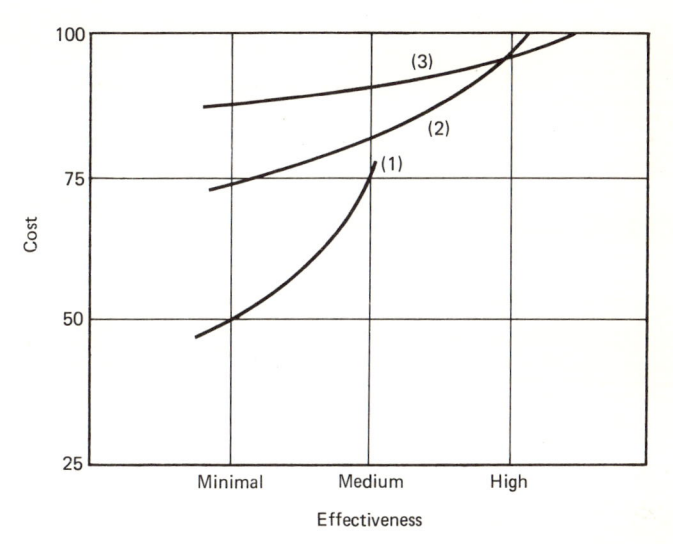

Figure 5-1 Relationship between cost and effectiveness for three instructional alternatives.

that the type of analysis illustrated above is somewhat unrealistic. We could proceed with such analysis if we had empirical data rather than subjective estimates on which to base the cost-effectiveness relationship; example 2 discusses a procedure based on such data.

EXAMPLE 2

The key to using cost-effectiveness analysis in education lies in the results of input-output-equations. These equations have two main advantages. (1) They include as inputs specific types of resources whose costs can be determined. (2) They permit estimating the marginal effect of a given treatment, with the effect of other inputs held constant.

The first plausible use of empirical data as a basis for cost effectiveness analysis applied to the production of education is that of Henry M. Levin.[3] Levin began with regression equations developed by Eric A. Hanushek, in a recent study at M.I.T.[4] On the basis of these equations, he calculated the "marginal effect of key input variables, that is, the effect on output of an increase of one unit of input." Since inputs associated with teachers' services are critical, he selected two of these for examination, "teachers' verbal score" and "teacher experience." On the basis of other data, he calculated the costs associated with increases in teachers' verbal scores and experience. These costs were then applied to the output data (Table 5-2).

Table 5-2 Relative Costs of Increasing Student Verbal Achievement

Strategy	Approximate Cost for Increasing a Student's Verbal Score by One Point	
	Negro	White
Teacher's Verbal Score	$ 26	$ 26
Teacher Experience	128	253

Source: Levin, op. cit., p. 12.

Table 5-2 corresponds to our definition of cost-effectiveness. The objective is to raise students' verbal scores by (on the average) one point. Two alternatives are available—to hire teachers with higher verbal scores or with more experience. The best solution from the cost-effectiveness point of view involves increasing "the recruitment and retention of verbally able teachers while paying somewhat less attention to experience."[5]

Additional empirical studies are needed, in order that input-output

---

[3] Henry M. Levin, "Cost-Effectiveness Analysis and Educational Policy—Profusion, Confusion, Promise." Research and Development Memorandum No. 41, Stanford Center for Research and Development in Teaching, School of Education, Stanford University, 1968, p. 9.

[4] Eric Hanushek, "The Education of Negroes and Whites." Unpublished Ph.D. Dissertation, Department of Economics, Massachusetts Institute of Technology, 1968.

[5] Levin, *op. cit.*, p. 12.

studies may be used as the basis of this kind of decision model. In particular, production functions need to be developed for various populations, using a variety of outcomes. Replication, at given intervals of time, will provide information about changes in the input-output relationship.

The work of Levin combines input-output and cost-effectiveness models. Both approaches are related to cost-benefit analysis, which, however, uses monetary returns as the output measure, and depends upon a longer period of benefit; increments in lifetime earnings are the basic output.

3. Cost-Benefit Models

The framework of the cost-benefit model was explicated in Chapter 2. The objective in this model is the difference between the present value of benefits and the present value of costs, *or* the internal rate of return. The alternatives consist of different ways of allocating resources, either between education and other public and private activities, *or* within the educational sector. The decision rules include the following:

(a) If the rate of return to investment in education is greater than that in the remainder of the economy, society should increase its expenditures for education.

(b) If the rate of return for one level of education (say, preschool, or secondary) is greater than that for another, resources should be re-allocated within the educational enterprise in favor of those levels with the higher rate of return.

(c) Investments in job-related programs, such as vocational education or manpower training programs should be determined in part by the present value of streams of income and cost resulting from these programs.

(d) An individual should invest in an increment in education under the following conditions: (i) if the present value of the additional income to be obtained is greater than the present value of costs (for whatever discount rate the individual chooses); or (ii) if the internal rate of return associated with obtaining a given amount of education is greater than the rate of return obtained from other investments which are considered.

There are, of course, limitations on a procedure which uses income as the criterion variable. Nevertheless, there are a number of kinds of decisions which can appropriately be based on this kind of analysis.

(a) Nations which are interested in increasing their rate of economic growth must include in their considerations the development of human capital. This involves carrying out a macroanalysis of the costs and benefits of education.[6]

(b) State departments of education which are evaluating the results of their vocational education programs must be concerned with the ability of graduates of the programs to obtain gainful employment. It is, therefore, appropriate that states carry out cost-benefit analysis of their vocational educational programs.[7]

(c) School boards as well as teachers may base decisions on a cost-benefit analysis of salary schedules. School boards' decisions in establishing the schedules will be affected by the effect they have on teachers' decisions to upgrade their qualifications. Teachers decisions to enter the profession, to upgrade their qualifications, or to move from one school system to another are probably affected by their perceptions of the salary benefits they will receive as a result of their action, as well as by the concomitant costs.[8]

In spite of its apparent clarity, the cost-benefit approach leaves some problems unsolved. In the first place, the outcomes of education include consumption as well as investment benefits. (Some analysts would include deferred consumption as part of the investment; like present consumption, this does not show up in the computation of monetary benefits.) To include consumption benefits would constitute an increase in the total benefits associated with investment in education, and would justify a higher level of expenditures than would be justified on the basis of income benefits alone. On the other hand, to the extent that the apparent return to investment in education reflects the additional salary reaped by those with greater "native potential," higher motivation, or higher parental status, the benefits of education may be overstated, and a smaller expenditure may be appropriate.

These problems do not justify discarding this important analytic

[6] There is a large and growing literature on the relationship between education and economic growth. See, for example, Samuel Bowles, *Planning Educational Systems for Economic Growth* (Cambridge, Mass. Harvard University Press, 1969).

[7] Among the recent studies is that of Corazzini, *op. cit.* Evaluations development training programs may be considered under the same rubric. These studies include Earl D. Main, "A Nationwide Evaluation of M.D.T.A. Institutional Job Training," *The Journal of Human Resources*, III, No. 2 (Spring, 1968) pp. 159–170

[8] See Harvey, *op. cit.*

procedure. Rather, the results of the procedure should be treated with some caution, and the assumptions behind cost-benefit analysis should also be kept in mind. This caevat holds true, also, when linear programming is considered.

Linear Programming

The discussion of the use of cost-benefit analysis for decision making would not be complete without an examination of linear programming. Programming (linear and nonlinear) is concerned with finding the best, or optimal solution to a given problem.[9] Used extensively by economists, it is a mathematical technique, which has application to a wide variety of problems. Three examples of problems which are appropriate for analysis by linear programming are the following.[10]

(a) *The determination of optimal product lines and production processes.* Business analysts have used linear programming to determine the amount of production of various products which would lead to the maximization of profit. They have also tried to determine optimal production processes through this technique. An application to education of this latter application will be discussed below.

(b) *The selection of transportation routes.* Attempts have been made to use linear programming to select optimal transportation routes for an industry, together with the best location for its factories and its storage facilities. There is an obvious application of this procedure to the planning of pupil transportation systems.[11]

(c) Linear programming techniques have also been used to determine the least cost combination of ingredients to provide for a product meeting a given set of specifications. If, for example, it is known that gasoline must contain at least a certain amount of three ingredients, linear programming will suggest ways in which blending may be carried out at least cost, in keeping with these specifications. These procedures have been used in planning school lunch menus.[12]

---

[9] William J. Baumol, *Economic Theory and Operations Analysis* (Englewood Cliffs, N. J.: Prentice-Hall, 1964), ch. 5.
[10] *Ibid.*, p. 65.
[11] C. West Churchman, Russell L. Ackoff, and E. Leonard Arnoff, *Introduction to Operations Research* (New York: Wiley, 1957), ch. 11.
[12] See *Ibid.*, for methodology involved.

There are three basic elements in a linear programming problem. The first is the identification of the variable which is to be optimized. In the case of a business firm, this variable may be total profit. In the case of a school system, it may be the number of pupils educated to a certain level, the academic achievement (for example, in reading) of a given group of students, or the difference between the present value of income and cost streams.

The second element consists of side conditions, usually expressed as inequalities. These are the limitations or constraints placed on the process. In the transportation problem these include the cargo carrying capacities of trains or boats. In the product mix problem, the constraints are the amounts of different ingredients which are required. In production problems, the constraints consist, in part, of resource limitations.

The third element in a linear programming model is less obvious. It is necessary, for computational purposes, to insist that all variables be non-negative.

The following model is illustrative of this process.[13]

1. In a given educational system, a team of planners is faced with the task of maximizing the total utility of an educational system consisting of n levels:

$$x_j = x_1, x_2, \ldots x_n$$

x_j represents the number of students completing educational levels $1 \ldots n$

$$U_j = U_1, U_2, \ldots U_n$$

U_j represents the utility (expressed, for example, as the present value of expected income minus costs) associated with each educational level.

It is desired to optimize total utility:

$$U_j x_j = U_1 x_1 + U_2 x_2 + \cdots U_n x_n$$

2. Certain limitations are imposed by the shortage of resources.

$$a_{ij} = a_{11}, a_{21}, \ldots a_{m1}, a_{12}, \ldots a_{mn}$$

The a_{ij} are the amounts of input b_i necessary to produce a graduate of level j under assumptions about minimal standards (teacher-pupil ratios, textbooks, physical plant, etc.).

$$b_i = b_1, b_2, \ldots b_m$$

[13] Russell G. Davis, *Planning Human Resource Development* (Chicago: Rand McNally, 1966), p. 151.

The b_i are the resources available for education (money available for capital outlay for plant, trained teachers, equipment, books, etc.). The amount of each resource used in each aspect of the productive system is limited by the total amount of that resource which is available.

$$a_{11}x_1 + a_{12}x_2 + \ldots + a_{1n}x_n \leq b_1$$
$$a_{21}x_1 + a_{22}x_2 + \ldots + a_{2n}x_n \leq b_2$$
$$a_{m1}x_1 + a_{m2}x_2 + \ldots + a_{mn}x_n \leq b_m$$

3. The solution does not permit solutions consisting of the production of negative numbers of graduates.

$$x_1 \geq 0, x_2 \geq 0, x_3 \geq 0 \ldots x_n \geq 0$$

Bowles used similar procedures to examine the allocation of resources for education in Northern Nigeria.[14] He used the present value of total expected income minus the present value of costs as his criterion variable. The constraints were similar to those described above. His model was addressed to important sociopolitical questions such as including the following:

(a) What amount of society's resources should be devoted to the educational system?
(b) How should the total resource use be distributed among various types of educational institutions?
(c) What educational technologies should be chosen?
(d) What is the optimal level and composition of the importation of educated labor?

His study, and the Argentine study of Irma Adelman[15] suggest that this procedure can be operationalized for use in education. Additional applications of this type of model to many types of educational problems may be anticipated in the future.

Economic Aspects of Curriculum Decisions

The selection of curricula and the division of time among subject areas, as well as the assignment of students on the basis of ability or other criteria to a given program of instruction can be subjected to

---

[14] S. S. Bowles, *The Efficient Allocation of Resources in Education: A Planning Model with Applications to Northern Nigeria,* Ph.D. dissertation submitted to Harvard University, 1965.

[15] Irma Adelman, "A Linear Programming Model of Educational Planning: A Case Study of Argentina," in *The Theory and Design of Educational Development,* edited by Irma Adelman and Erick Thorbecke (Baltimore: Johns Hopkins Press, 1966).

cost-benefit analysis. However, much of this analysis must of necessity be loose and imprecise. For example, it would be inappropriate to decide, on the basis of monetary costs and benefits, whether or not world history should be taught in the ninth grade. The use of cost analysis in curricular decisions may result in ludicrous conclusions, as Callahan[16] pointed out.

There are, however, certain curricula for which cost-benefit analysis is an appropriate decision making technique. These are the vocational and professional programs,[17] which are specifically designed to prepare students for employment. Here, future occupational success is an appropriate measure of output, while the cost of training is an important input measure. If two programs at the same level (for example, for Grade II boys) are compared, foregone earnings can be ignored and the differences between the costs and benefits of the two programs (for example, automotive mechanics and electronics) may be examined.

1. Costs. Vocational education programs are more costly than general academic programs. The additional cost is due to such factors as (a) the high salaries which vocational teachers can often command, due to their relative short supply; (b) the small classes often mandated for safety reasons in vocational education; and (c) the relatively high implicit rent of space and equipment, which results from the need for larger rooms, and for expensive machines. A recent study compares the cost of vocational education with that of a nonvocational high school (Table 5-3).

2. Benefits. Some data on original entry wages for graduates of vocational and other high school programs are now becoming available. These data must be interpreted with care, since students are selected for various high school programs on the basis of ability, interests, and motivations, and their subsequent job experience will reflect these variables in addition to the nature of their training. Data now available suggest that vocational school graduates can command a higher wage on original entry into the labor market than other graduates, although the effect on this difference of corrections for ability is not yet known. Corazzini[18] has compared entry wages, and has calculated the number

---

[16] Raymond E. Callahan, *op. cit.*

[17] Milton Friedman and Simon Kuznets, *Income from Independent Professional Practice* (New York: National Bureau of Economic Research, No. 45, 1945).

[18] Arthur J. Corazzini, "The Decision to Invest in Vocational Education: An Analyses of Costs and Benefits," *Journal of Human Resources,* **III,** (Supplement, 1968), pp. 104, 106. See also Edward S. Todd, "High School Program of Study and Subsequent Employment of Terminal Graduates," Ph.D. dissertation, Department of Education, University of Chicago.

Table 5-3 Total Resource Costs, Worcester Public High School, and Worcester Boys' Vocational School, 1963–1964*

	Public High School ($ per pupil)	Boys' Vocational High School ($ per pupil)
Total public cost	532	1210
Current cost	452	964
Implicit rent†	59	165
Property tax loss	21	81
Total private costs	1176	1176
School related costs§	56	56
Foregone earnings	1120	1120
Total resource costs	1708	2386

Sources: Worcester Public Schools, Office of the Superintendent; Worcester Boys' Vocational High School and Worcester Industrial Technical Institute; Office of the Assessor, City of Worcester; and Massachusetts State Department of Education, Department of School Building Assistance.

* Arthur J. Corazzini, "The Decision to Invest in Vocational Education: An Analysis of Costs and Benefits," *Journal of Human Resources,* III (Supplement, 1968), p. 102.

† Implicit rent and property tax losses estimated using techniques in Theodore W. Schultz, "Capital Formation by Education," *Journal of Political Economy,* LXVIII (December, 1960), p. 575; and Fritz Machlup, *The Production and Distribution of Knowledge in the United States* (Princeton, N. J.: Princeton University Press, 1962), p. 100.

§ Private costs estimated using techniques in Schultz, *op. cit.,* p. 575.

of years these initial differentials would have to persist in order that their present value would equal the present value of the added costs (Tables 5-4 and 5-5).

Private Decisions to Invest in Education

There are important limitations on the utility of cost-benefit analysis for private decision making.

(a) The data used in the computation of present value and rate of return are based on the costs and earnings of large numbers of individuals, and the resulting calculations are averages, which

Table 5-4 Starting Wages Paid Vocational High School and Regular High School Graduates Who Were Hired as Machine Operators in Worcester, June 1964 to June 1965

Total Number of Employees in Firm Employing Trainees	Number of Firms Sampled	Average Starting Wage High School Graduate	Average Starting Wage Vocational School Graduate	Average Differential in Starting Wage
More than 1000	2	$1.95/hour	$1.99/hour	$0.00/hour
More than 500, less than 1000	3	1.76/hour	1.89/hour	0.13/hour
More than 200, less than 500	2	1.64/hour	1.82/hour	0.18/hour
More than 100, less than 200	3	1.70/hour	1.98/hour	0.28/hour
More than 25, less than 100	2	1.70/hour	1.95/hour	0.25/hour

Source: Corazzini, op. cit., p. 104.

Table 5-5 Number of Years Wage Differentials Would Have to Remain in Order for the Present Value of Extra Costs to Be Equal to the Present Value of Extra Returns

Wage Differential	Rate of 5% Discount	Rate of 10% Discount
$ 80	Never equated	Never equated
260	17	Never equated
360	11	30
560	6⅙	10
500	7	12

Source: Corazzini, op. cit., p. 106.

no doubt conceal wide variations. Each individual is therefore faced with uncertainty as to the results of his decision, and he must make a subjective judgment as to his position on the income distribution of individuals with his level of education. On the other hand, the individual has means of removing this type of limitation. He has knowledge, albeit subjective, about his own ability which is not available to others. He can make instant use of feedback and since most decisions are subject to

modification, he can remove a considerable portion of this uncertainty.
(b) Changes in society and the economy will require new kinds of skills in the future. This introduces uncertainty into students' present decisions to invest in their education.
(c) The individual faces the possibility that his investment will not pay off, due to poor health or even premature death.
(d) An individual may be unable to finance his own education at a reasonable rate of interest.

To what degree are individual decisions to invest in education wise or rational? A private decision to take additional schooling is more likely to be a rational one during a period when the demand for college and high school graduates is increasing, and when the rates of return to investment in education are well above the rates obtainable for other investment. Nevertheless, since cost and benefit data are often almost nonexistent, and are seldom made available to private decision makers, it can be questioned whether individuals' choices are in fact influenced by rational considerations. Additional research, based on the relationship between the cost-benefit calculus and private decision-making behavior is needed.

PROBABILITY MODELS

Whether he makes his decisions as a result of input-output analysis, or on a more intuitive basis, the administrator must devise a strategy for his decision making. This strategy will include a subjective evaluation of the result of accepting one course of action rather than another. Such evaluations may tend to be implicit; however, an administrator who involves others in his decisions will be required to state the assumptions of the proposed courses of action in more explicit terms. The following example, based on the literature of game theory and statistical decision making, formalizes the decision making process.

The example is stimulated by the work of Chernoff and Moses, and Luce and Raiffa, *op. cit.*

EXAMPLE 3

Consider the case of a superintendent who faces the task of hiring teachers.

In the case of each applicant, we assume that he has only two possible choices:

a_1 hire
a_2 do not hire

The superintendent does not know what the outcome of his decision will be. Let us call this outcome the "state of nature." There are three forms, in this example, which the state of nature can take, namely:

θ_1 The teacher will be successful
θ_2 The teacher will be mediocre, but acceptable
θ_3 The teacher will be a complete fizzle

In thinking over his problem, the superintendent decides that there are some economic losses associated with each possible decision (Table 5-6). For example, his hiring of a mediocre teacher will result in a poor use of money because students will do poorly. On the other hand, the losses incurred as a result of hiring a very poor teacher may be very great. Let us assume that our superintendent develops a probability table in which he lists his estimate of the loss which is associated with each possible decision. Note that even the decision to hire a teacher who proves to be successful incurs some loss—namely, the cost of the hiring procedure. Losses associated with decisions *not* to hire result from costs of obtaining information, and the loss to the district of not obtaining the services of a teacher. For example, in an extreme case, the teacher may be dismissed but his contract may require that his salary still be paid.

Table 5-6   Loss Table, as Estimated by Superintendent

Actions States of Nature		a_1	a_2
	θ_1	1	6
	θ_2	6	3
	θ_3	8	2

Our superintendent is now faced with the necessity of formulating a hiring strategy. He could adopt such strategies as: hire only teachers with relatives on the board; hire everyone (if his pool of applicants is limited); hire on the basis of the toss of a coin (if his pool is about twice the number of vacancies) and so on. However, he discards these possibilities, and instead decides to rely on the results of a screening instrument his research director has developed. This instrument is not infallible, but in a series of past experiences, its correspondence to the states of nature has been established (Table 5-7).

Table 5-7 Predictions of Success from Screening Instrument (Scored on Ten-Point Scale)

Observations in Terms of Range of Scores		z_1 (7–10)	z_2 (4–6)	z_3 (0–3)
States of Nature	θ_1	3/5	2/5	0
	θ_2	1/5	3/5	1/5
	θ_3	0	2/5	3/5

Table 5-7 should be interpreted as follows. Three-fifths of the potentially successful teachers score in the z_1 range, and two-fifths in the z_2 range.

The superintendent is in a better position, now, to formulate his alternatives. He may, for example, hire all teachers who are in the z_1 range, and not hire any others, even though he knows he will miss some good bets who happen to score z_2. At any rate, we are now in a position to list all possible strategies based on the z's. Furthermore, it is possible to compare the results of these strategies in terms of the losses which might be anticipated in each case.

Table 5-8 A Listing of Possible Strategies

Strategies Observations	s_1	s_2	s_3	s_4	s_5	s_6	s_7	s_8
z_1	a_1	a_1	a_1	a_1	a_2	a_2	a_2	a_2
z_2	a_1	a_1	a_2	a_2	a_1	a_1	a_2	a_2
z_3	a_1	a_2	a_1	a_2	a_1	a_2	a_1	a_2

Table 5-8 should be read as follows: strategy 2 (s_2) is to take action a_1 (hire) if the applicant's score is z_1 or z_2 and action a_2 (do not hire) if the score is z_3. (Strategy 1 calls for hiring all applicants, and strategy 8, equally drastic, says "hire no one.")

Finally, it is possible to evaluate and compare our strategies on the basis of the loss table. The average losses associated with the various strategies and the three states of nature is shown in Table 5-9.

Table 5-9   Losses Associated with Strategies

Strategies State of Nature	s_1	s_2	s_3	s_4	s_5	s_6	s_7	s_8
θ_1	1	1.0	3.0	3.0	3.0	4.0	6.0	6.0
θ_2	6	5.4	3.6	3.6	5.4	4.8	3.6	3.0
θ_3	8	4.4	5.6	2.0	8.0	4.4	5.6	2.0

The method of arriving at the figures in Table 5-9 is as follows: Consider strategy 2 (hire if z_1 or z_2; do not hire if z_3). Loss if state of nature is θ_1 is $3/5(1) + 2/5(1) + 0(6) = 1.0$. Loss if state of nature is θ_2 is $1/5(6) + 3/5(6) + 1/5(3) = 5.4$, and so on.

Notice that strategy s_4 which says, hire if the individual scores in the z_1 range, otherwise do not, seems best. It is, however, only better than s_8 (hire no one) if the state of nature is θ_1. That is, the result of this hypothetical analysis, so far, is that a conservative policy toward hiring is desirable. To be sure the results would change if the "loss table" were altered.

Finally, our decision maker is better off if he knows something about the distribution of the state of nature. Suppose, for example, that on the basis of previous experience he knows that, on the average, 40 percent of teachers are successful, 40 percent are mediocre, and 20 percent are fizzles. Then, he can for each strategy, compute the *average expected loss*. In the case of strategies 4 and 8, this is:

Strategy 4: $3(.4) + 3.6(.4) + 2(.2) = 3.04$
Strategy 8: $6(.4) + 3.0(.4) + 2(.2) = 4.0$

Strategy 4 is now clearly the best. Or is it?

FEEDBACK MODELS

Open systems rely on feedback mechanisms to permit them to adapt to changes in the environment. These mechanisms and adaptive procedures vary. Biological systems are extremely sensitive to environmental change. The human body, for example, is able to remain at a temperature in the neighborhood of 98°F, in spite of wide variations in the temperature of the surrounding air.

Social systems, on the other hand, rely on feedback mechanisms which are less precise. As a result, human organizations often fail to adapt to a changing environment. If the organizations are given a guaranteed source of resources (as is the case with tax-supported institutions), the motivation to respond to the environment may be removed, and the lag may increase. Current demands for "accountability" may be interpreted as calls for the more effective use of feedback.

The environment of educational systems is constantly changing. For example, technological change may call for a different type of educational output, in the form of students who can take new types of employment, and who can adapt to a changing world. Schools in changing neighborhoods may be faced with a new type of input, in the form of students with different kinds of home backgrounds.

What kinds of feedback mechanisms do school systems possess? How useful are these mechanisms in helping these systems to adapt to external change? Are schools sufficiently adaptive, or would public schools be unable to compete with privately supported institutions on a cost-benefit basis?

Schools and other organizations possess both formal and informal mechanisms for adapting to change. The formal mechanisms include the organization's rules for responding to given stimulae. For example, rules for closing schools because of severe weather or epidemics will be found in many systems.

Informal procedures are also extremely important. Schools, like other organizations, possess an informal organization as well as the formal organization outlined by charts and the written rules of the system. The informal organization provides ways for the organization to adapt to pressures without calling its cumbersome decision making machinery into play. Frequently, for example, internal conflict among teachers or groups of teachers may be mediated by the informal organizations, while the formally appointed decision makers may know little or nothing about the existence of the conflict or of its nature. Equally important, the informal organization of students appears to mediate the instructional process, and to help determine which organizational values will receive priority.

The task of responding to feedback from the environment cannot be left entirely to the informal organization. In order that the formal organization may accomplish its goals, it must compare feedback from its environment with its own progress. Indeed, it must also, at times, attempt to influence the environment.

There is some need for feedback mechanism and decision rules that are more or less automatic, and do not require that formal decisions be made each time feedback is to be used, but that imply direction and movement as well as adaptability.

One such mechanism is geared to the maintenance of organizational productivity. The important parts of such a feedback system are the following: (1) Procedures for measuring output, at various stages. This includes a monitoring of performance at the various levels of the educational process, and also the monitoring of post-school success—in obtaining employment, earning an income, participating in political life, and so on. (2) Procedures for ordering these findings, and comparing them to the desired level of performance. (3) Procedures for using the difference between actual and desired performance to affect inputs and processes.

In a highly efficient educational system, feedback would be continuous and immediate. The results of the feedback would govern the activities within the organization. Figure 5-2 illustrates the use of feedback in a simple educational system containing only one teacher and one student.[19]

Equation (3) in Figure 5-2 is the "theory" upon which the system

[19] Adapted from Herbert A. Simon, "Application of Servomechanism Theory to Production Control," in *Models of Man* (New York: Wiley and Sons, 1957), pp. 219–240.

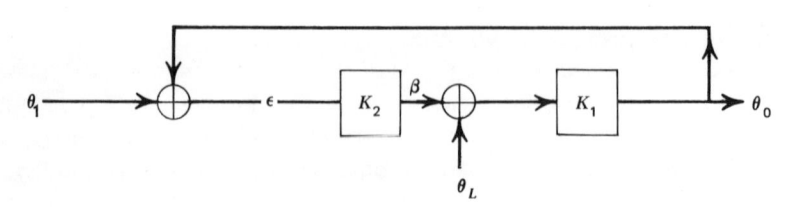

Figure 5-2 A one-teacher, one-student educational system.

θ_I = desired level of performance
θ_0 = actual level of performance
θ_L = load (capacity of student)

Equations:
(1) $\epsilon = \theta_I - \theta_0$ (error is difference between actual and desired level of performance)
(2) β = "intensity" of instruction
(3) $\theta_0(t) - \theta_0(t-1) = K_2[\beta(t) - \theta_L(t)]$
(4) $\beta(t) = K_2[\epsilon(t)]$
(5) $\epsilon(t) = \theta_I(t) - \theta_0(t)$

is built. It states that the performance increment in a given time period is a function of instruction β and the capacity of the student θ_L. (This latter factor includes native ability and other background effects, but also the level of performance which the student had reached at time $t\text{-}1$).

Equation (4) is the decision rule. By means of the feedback process, actual performance is compared with desired performance, and the error term is calculated. The decision rule relates the error (difference between desired and actual performance) to the type of instruction. The implicit idea is that the educational environment does vary in intensity, and that this intensity can be adjusted to the magnitude of the error term. For example, if the performance is close to the desired level, formal instruction can be reduced, and the student can be encouraged to do additional reading. If the error term is large, intensive instruction (for example, additional help, more emotional support, additional time of student and teacher) may be required. In the design of a given system or of a simulation model, the decision rule would have to be made much more explicit.

An important part of the total process is the way in which expectations are formed, since the expectations of teachers, parents, and students will help to determine the performance levels which are reached, just as the "setting" of the thermostat ultimately governs the temperature of a room: Empirical studies of teachers' and administrators' expectations for student performance and the manner in which these expectations are developed are necessary if efficient educational systems are to be developed.

Any system must have adequate information in order to operate properly. A thermostat depends for its operation upon its ability to acquire information about present temperature of a room. Living *organisms* depend on sensory devices to provide the data which make self-regulation possible. For example, temperature control mechanisms (such as pores which open and close) depend upon accurate readings of the body temperature.

Organizations also need large quantities of information. Business firms make decisions based on information about the estimated market demand for their products at various selling prices, about the present level of inventory, and about the cost of production. In efficient firms, a good deal of effort and expense goes into making this information available. Market research teams may spend years studying the potential demand for a promised product before the decision is made to manufacture it.

Much of the information gathering which characterizes large secondary schools is less formal and probably less effective than that carried out by businesses. For example, the school principal rarely conducts intensive "market surveys" of his attendance area, although the information which he might obtain in this way about demand for education could be extremely useful for the development of curricula and the determination of expectation levels for education in the community.

As noted earlier, information about the costs and benefits associated with alternative instructional or organizational procedures is necessary for improved decision making. It is important that such data be gathered systematically, so that decisions in education may be based on data and not on current fads and fashions.

Teachers need information about students, to use as a basis for making decisions about teaching procedures. It is often very difficult for teachers to obtain this information. Teachers who have to deal with as many as 150 students in the course of a day may know relatively little about their students, and hence may be unable to provide a personalized direction of learning.

The above examples illustrate the place which information plays in the regulation of an educational system. There must be "receptors" or information collecting mechanisms, which gather the data which the system needs in order to operate effectively. These mechanisms may be built into the organization by defining appropriate roles and positions. For example, there may be people charged with gathering information about the community, about opportunities in the labor market and in institutions of higher education; about teacher-student, teacher-teacher, and student-student relationships; and about students' homes and other out-of-school influences.

The information must be transmitted to appropriate decision makers who are charged with formulating plans of action. At present, the task of analyzing and transmitting information is probably even less effectively performed than the task of initial data collection. Specially trained individuals are needed at all educational levels to permit the effective transfer of data, and its use in decision making.

The decision maker translates information into action. In the thermostat, information about the actual room temperature is compared with the expectation or desired temperature. If the difference is above a given tolerance level, the electronic decision maker switches on the furnace. This is an automatic decision, requiring no conscious choice. Automation in industry has increased the number of decisions which are made automatically.

It has been suggested that many positions in middle management can be automated. Decision rules can be established, and the computer can be instructed to make a decision, according to the information which has been fed into it. "Progress" has consequently been made toward the automation of production and distribution processes in the petroleum industry.

Some attempts at automation in education have been made through the medium of programmed learning, including the so-called "teaching machine." Students' responses are compared with the desired or correct response, and instructions are given to the student on the basis of the correspondence or difference which is discovered. Other suggestions have been made which would result in the establishment of an automated classroom.[20] However, the learning of all but the lower order cognitive processes is so subtle that only a sensitive and competent teacher can make the types of decisions which are needed to direct students' learning.

At the organization level, likewise, we are probably far from the automation of decision making; in fact, there is too little use of data by decision makers. Let us, for example, consider the problem of allocating a student's time according to the educational needs which are manifested. Suppose it is desired to provide each student with the educational activities which are appropriate, in accordance with the difference between his actual and expected performance.

Let ϵ be the difference between actual and expected performance.

Decision Rule

If $\epsilon > 10$ Select Activity 1
If $5 < \epsilon < 10$ Select Activity 2
If $\epsilon < 5$ Select Activity 3

Here

Activity 1 is: Provide small group and individual instruction
Activity 2 is: Continue present class instruction
Activity 3 is: Excuse for library assignment

This decision rule is obviously oversimplified. However, an organization operated on this basis could be used to provide a more individualized type of instruction than most schools currently offer.

INFORMATION SYSTEMS

Large amounts of reliable information are essential for making sound decisions. In terms of the models presented in this chapter, data about

---

[20] See, for example, Simon Ramo, "A New Technique for Education" in *Automation: Implications for The Future* (New York: Random House, 1962), pp. 428–444.

inputs and outputs constitute the basis on which choices are made. The feedback models discussed in the last section also require a steady flow of information in order to operate efficiently.

The need for information is an essential aspect of open systems theory. Systems exist in constant interaction with their environment. The environment is in a state of change, and systems which do not possess information retrieving procedures are not able to adapt to environmental change.

At present, data sources at the state and local level tend, at best, to be inadequate. State legislatures often are required to make decisions involving hundreds of millions of dollars, in the absence of the most rudimentary information about programs and about educational outcomes. Even local school systems, have an inadequate flow of data, and school boards must base their choices on judgments of administrators, often unsupported by sufficient information. Efforts of school board members to obtain staff to help them gather and analyze data may sometimes be resisted by the administrative hierarchy.[21]

Where data exist, they are often not in the form which is required for the improvement of decision making. Data tend to be fragmented, and to become the property of the various subsystems. For example, there are data about pupil guidance, achievement, and finance, but information about resource allocation and achievement are rarely used jointly as a basis for making decisions about the distribution of educational opportunity.

Serious efforts are being made to develop classifications of educational data and means of obtaining and storing data that are compatible among states. However, these vast taxonomies do not necessarily provide the basis for the improvement of decisions. It is entirely possible that the costs of obtaining this information will exceed the benefits, unless it is consciously and efficiently used for the improvement of decision making. As additional money is spent on acquiring data, analysis of the costs and benefits associated with gathering these data should be conducted. Procedures for minimizing cost of acquiring usable data, including sharing of data among school systems and between school systems and other units of government, should be developed.[22]

In order that additional information may be functional, it must be conceptualized in such a form as to permit its use in decision making. The classification of data is discussed next.

[21] David Rogers, *110 Livingston Street* (New York: Random House, 1968).
[22] See George J. Stigler, "Information in The Labor Market." *Journal of Political Economy* LXX, No. 5, Part 2 (Supplement: October, 1962), pp. 94–105.

Categories of Relevant Data

The types of information systems which should be used vary by organizational level. Data used by state educational agencies should be compatible with locally developed systems, but there are differences, and the two levels should be examined separately.

1. State Agencies

State agencies have some legal responsibility with respect to inputs into educational systems. The following are examples:

(a) Teacher inputs. State agencies typically certify teachers. This provides a state level quality control over members of the teaching profession. Often, certification is also required of administrators, counsellors, and other personnel. Hence, state studies of teacher qualifications, teacher mobility, and attrition are appropriate. These data can form the basis for state studies of the relationships between teacher qualifications, teachers' salaries, and educational outcomes.
(b) School plant. State agencies also assume responsibility for the supervision of school facilities, particularly with respect to their safety.
(c) Other inputs. At this date, states do not typically supervise such inputs as educational television and language laboratories. In some cases they may suggest lists of books for school libraries.
(d) Student inputs. In order that states may ensure suitable educational opportunities for all students, some state level data concerning student inputs is also desirable.

There tends to be little information available at the state level about the educational programs offered by school systems. However, as state legislators are required to invest additional millions of dollars for the support of education, they are demanding more information about the programs supported with this money.

States tend to gather very little data about outputs. Yet, feedback mechanisms linking financial inputs to outputs are almost essential if equal educational opportunity is to be promoted. This suggests a need for statewide testing. It also suggests a need for the monitoring of the post-school employment and educational experience of students, at least on a sample basis.

2. Local Agencies

In general, the data required by local educational agencies is more detailed and more specific than that collected at the state level. Although state input-output studies may be related mostly to fiscal procedures, local studies should be directed to the improvement of educational practice, through the promotion of changes in instruction and organization.

- (a) *Input data.* Specific student input data will be needed at the local level, including some information about the educative influences in students' homes. Teacher data would, if possible, include measures of attitude, as well as detailed information about qualifications and experience. Again, detailed data about school facilities are required at the local level.
- (b) *Output data.* Achievement and follow-up studies are even more important at this level. It is especially necessary that these data be linked, through feedback procedures, to the improvement of practice.

Intersystem Cooperation

To this point, educational systems have been viewed as separate, autonomous agencies. In actuality, although each may be independent in some respects, they are related in terms of their place in the total society. Local educational systems are integral parts of state systems. Also, while each state educational system is a legal entity, all are part of a *de facto* national system. Even at the national level, there are informal ties with educational systems in other nations; these ties are strengthened through UNESCO and similar agencies.

There are a number of barriers which inhibit the transfer of information among states. Some of these barriers are political, and are related to the supremacy of one or another party in each state. Furthermore, state legislatures are jealous of their prerogatives and are not favorable to the cooperation among states. Members of the state bureaucracies often have rather limited professional experience outside their own state.[23]

---

[23] J. Alan Thomas, "Governmental Cooperation to Improve Efficiency in Education," in *The School and the Challenge of Innovation* (New York: McGraw Hill, and the Committee for Educational Development, 1969).

In spite of these barriers, there are a number of reasons why interstate cooperation and interchange of data is desirable. Practices which have proven to be effective in one state may have substantial value in another. The interchange of data about production functions can lead to improved efficiency in each state, and in the nation as a whole.

The argument for interchange of data among local systems is even stronger. Many local systems do not have the resources which they would need, for example, to examine input-output combinations, and to identify promising educational strategies. When such information is produced in one system, it is highly desirable that it be made available elsewhere. State agencies share the responsibility for making possible this type of interchange of information.

Chapter Six

◆◇◆◇◆◇◆◇◆◇◆◇

Allocation and Budgeting

INTRODUCTION

Educational administrators have the responsibility of ensuring that their institutions meet the demands placed on them by society. They have a limited amount of resources at their disposal, and they must allocate these resources so as to maximize organizational productivity.[1] The main vehicle within which allocatory decisions are made is the budget.

Two central facts guide and motivate this analysis. In the first place, education is a valued "commodity" in our society. It is of value in and of itself, since there are certain intrinsic satisfactions which result from its "possession." It is also important as a means to other valued goals. From a societal point of view, education is seen as one avenue

---

[1] In this respect, the concern of the administrator coincides with that of the economist:
> Economics is the study of how man and society choose, with or without the use of money, to employ *scarce* resources, which could have alternative uses, to produce various commodities over time and distribute them for consumption, now and in the future, among various people and groups in society. [Paul A. Samuelson, *Economics: An Introductory Analysis* (New York: McGraw Hill, Seventh Edition, 1967), p. 5.

toward economic growth, national defense, the reduction of unemployment, and the preservation of political democracy. From the individual's perspective, education makes possible a higher income, more choices among occupations, and a more prestigious social position.

In the second place, although the demand for education is both high and expanding, every society and most individuals find that there are obstacles which prevent the demand from being fully met. Our society could, if it so desired, spend more—indeed, much more—for education. Hence, resources are not limited in an absolute sense, within the range of expenditure levels that would normally be considered. However, resources always have alternative uses; hence, resources are scarce by definition, since their utilization for one purpose prevents their being used to support other public or private activities.

Since the demand for education is great and the resources which are available for producing education are (by definition) limited, every attempt must be made to improve the efficiency with which education is produced. This means that various methods of production must be considered, and that those procedures should be chosen which are most successful in producing the desired objectives, within the limits imposed by resource limitations.

The central concept is "allocation." Efficiency requires that different ways of using such scarce resources as money, teachers' time, space, and students' time must be considered, so that the final choice may result in resources being used in an optimal time. Even the time of the school administrator—who is in turn responsible for the allocation of the resources at his disposal—is an important and very scarce resource.

The model used as a basis for this analysis has already been discussed in Chapter 5. Linear programming, which is not only a mathematical procedure, but also the basis for a logical analysis of this type of problem, underlies the following discussion.

APPLICATIONS

These procedures are very general, and have a wide variety of possible applications. Consider the following examples:

1. It is desired to allocate resources among a number of public and private investments, including education. The allocation problem is to determine what level of expenditure for various purposes,

including education, will result in the highest level of social well-being.[2]

2. A state or nation wishes to decide how to allocate a given amount of resources among its various levels of education. The criterion, again, may be the maximization of the total well being of the society.[3]

3. A school district wishes to consider alternative ways of "producing" its graduates. Goals are identified in terms of such criteria as success in college, employability, and achievement, both academic and social. A number of methods of reaching these goals are considered. These procedures may include traditional instructional and organizational patterns, as well as novel combinations of inputs, such as different uses of libraries, the use of modern technology, and differentiated staffing patterns.

4. A different type of problem, but one which is important at all levels of education, concerns the allocation of resources among curricula. State Boards of Education, faced with decisions concerning the approval of a new school of medicine, or local school boards which are considering expanding their vocational education offerings provide two examples of organizations grappling with this type of problem. Each organizational level may base its decisions on studies of anticipated costs and benefits. State and national agencies are probably in a better position to assess costs and benefits and to carry out a careful analysis based on societal well-being, while local agencies are affected to a greater extent by demands expressed by parents and other citizens.[4]

5. A large city school board allocates $500,000 for the improvement of reading instruction in its ghetto schools. The problem is defined as allocating resources so as to maximize the attainment of its

---

[2] Most of the work in the Economics of Education is relevant to this question. Consider, for example, Theodore W. Schultz, *The Economic Value of Education*. (New York: Columbia University Press, 1963); Gary S. Becker, *Human Capital* (New York: Columbia University Press, 1964). Also, Samuel S. Bowles, "The Efficient Allocation of Resources in Education: A Planning Model with Applications to Northern Nigeria," Ph.D. dissertation, Harvard University, 1965.

[3] Bowles, *op. cit.*

[4] For examples of cost-benefit analysis of curricular issues, see Arthur J. Corazzini, "The Decision to Invest in Vocational Education: An Analysis of Costs and Benefits," *The Journal of Human Resources,* III (Supplement, 1968), pp. 88–120. Note that the Congressional decision to support the improvement of education in science, mathematics, and foreign languages reflected a federal concern for the societal benefits of these courses.

objective, improved achievement in reading. Possible alternatives may include reductions in class size, hiring of specially trained teachers, the purchase of additional books, or bussing students to integrated schools.[5] One long term solution may be to expand the system of early childhood education.

6. A school principal is faced with the task of constructing a schedule of classes for teachers and students. Here, the problem is one of allocating time rather than money or goods and services. This time allocation problem has been given remarkably little attention, especially at a theoretical level. One of the greatest contributions of economists to the internal efficiency of educational systems is their emphasis on the importance of students' time. Students' time is valuable regardless of foregone earnings, because of the opportunity costs involved when one activity is engaged in at the expense of other feasible alternative activities.

Each of these examples illustrates a problem in resource allocation. In order that these problems may be solved, information about input-output relationships is required.

ALLOCATION AND INPUT-OUTPUT ANALYSIS

Typical Allocatory Procedures

Present procedures for resource allocation in education are largely based on common practice and the judgments of authorities. In most cases, these procedures imply the use of ratios or indices to determine the inputs required to provide schooling for a given number of students. These indices include teacher-pupil and counselor-pupil ratios, per pupil square foot requirements for space, and recommended numbers of books per pupil for the school library. Some of these ratios result from the advice of experts in the field of public health (for example, the sanitary requirements of a school of given size) or safety (for example, the specification of building materials, emergency exits, fire extinguishers, or the construction of school buses). Some are based on traditional practice, rather than research.

[5] See David K. Cohen, "Policy for the Public Schools: Compensation on Integration," prepared for the U. S. Commission on Civil Rights, November 16–18, 1967, Conference on Educational Opportunity.

Table 6-1 California Building-Aid Law

School	Enrollment	Square feet per pupil
Elementary school comprising kindergarten and grades 1 to 6, inclusive	300 or more	55
Elementary school comprising grades 7 and 8	750 or more	75
Junior high school comprising grades 7 to 9, inclusive	750 or more	75
Junior high school comprising grades 7 to 10, inclusive	750 or more	75
High school comprising grades 7 to 12, inclusive	750 or more	80
Junior college comprising grades 11 to 14, inclusive	750 or more	80

Source: Wallace H. Strevell and Arvid J. Burke, *Administration of The School Building Program* (New York: McGraw-Hill, 1959), p. 153.

At times, these indices become imbedded in law. Table 6-1 shows the basis on which the State of California assisted school districts in capital construction in 1951.

The California legal requirement is an example of the way in which common practice becomes the basis for prescriptions for administration. Many of the indices published in educational textbooks are stated in terms of what "should be." Unfortunately, these prescriptive statements are often based on very inadequate empirical data about cause-effect relationships in education. Since conditions, including knowledge about the best ways to produce different kinds of education for different kinds of students, are in constant flux, this tendency to transform descriptive statements into prescriptions should be resisted. The recent incorporation of precise class size requirements in some salary agreements is an example of a formalization of current practice which is intended to improve working conditions and educational output, but which may in the long run retard experimentation with various input combinations, and hence impede the improvement of productivity. For this reason, these classes may be to the financial disadvantage of teachers, as well as to the detriment of the educational program.

One result of this uncritical acceptance of input indices is that, when additional resources become available, they tend to be distributed in roughly the same proportion as existing expenditures. This tendency may result from the desire of each subsystem to expand and improve its position with respect to the total organization. It has been suggested that the management system of an organization is particularly likely to add to its ranks. Members of the administrative hierarchy may add to their staff as a way of improving their status; however, increases in central office school administrators will not necessarily increase productivity.[6]

The correct method of increasing productivity is to allocate additional resources to those inputs with the highest payoff probabilities. For example, studies reported elsewhere in this book suggest that teachers' experience and verbal ability may be important contributors to output. The implication is that some reallocation of resources to improve teacher inputs will probably result in improved outputs in school systems resembling those upon which the statistical studies are based.

Significant though these studies are for the improvement of practice, there are some important reservations to be made.

1. Evidence based on cross-sectional studies made at a given point in time cannot always be used to guide policy making for the future. The variables which are statistically significant at one time may not be important predictors of output in the future, especially if other important variables are subject to change. This fact calls for conducting a series of regression studies, over a period of years, to permit an observation of changes in the regression coefficients.
2. The goal of input-output studies is to obtain estimates of the marginal effects of an increase in a given input variable. For example, we may wish to learn the effects on outputs of increasing the cost of teachers' time (for example, by higher salaries and in-service education) by, say, 10 percent.
3. Some of the variables with which we are dealing are probably proxies for underlying variables. Beginning teachers' salaries can best be seen as affecting outputs if they are viewed as part of a rational process for identifying and recruiting teachers for a given school system. To change allocations merely on the face value of the variables used in the research (often selected because data is available for these variables and not for others) would be to mistake the shadow for the substance.

Cost-Benefit Analysis and Allocation

Cost-benefit analysis is directed more specifically at allocatory problems, since inputs are expressed in monetary terms. Where this kind

[6] Daniel Katz and Robert L. Kahn, *The Social Psychology of Organizations* (New York: Wiley, 1966), p. 99. Also see C. N. Parkinson, *Parkinson's Law* (Boston: Houghton-Mifflin, 1957).

of analysis can be operationalized, the computed results (expressed as a rate of return) can be compared in strict numerical terms, and conclusions can be readily formulated.

Take for example the following statistics (Table 6-2).

Table 6-2   Estimates of Private Rates of Return, United States (%)

1. High school graduates, white males, after personal taxes 1958	28
2. College graduates, white males, after personal taxes, 1958	14.8
3. U.S. private domestic economy: implicit rate of return after profit taxes but before personal taxes 1957–1958	12.3

From Theodore W. Schultz, "Resources for Higher Education: An Economist's View," *The Journal of Political Economy*, LXXVI, No. 3 (May/June, 1968), p. 377.

The implication of Table 6-2 is that, since investment in secondary education brings such a high rate of return, society should investment more of its resources at this level. Higher education is also a good investment, but it does not warrant as much additional resources as education at the secondary school level.

Such a conclusion is, of course, rich in policy implications. However, cost-benefit analysis should be used with some reservations, as a guide to the formulation of policy, for the following reasons:[7]

(a) These studies are based on the assumption that earnings reflect social productivity. That is, it is assumed that what people are paid is a good measure of what they contribute to society. There are a number of indications that this is not always the case. For example, trade unions and professional organizations often pay individuals with similar previous experience and paper qualifications at the same rate, regardless of the quality of services they provide and, therefore, regardless of their contribution to organizational productivity. This phenomenon is very evident in education; the single salary schedule is prevalent in most of the country, and "merit" differentials are anathema to most teacher organizations.[8]

[7] For a discussion of the limitations of rate of return analysis in educational planning, see George Psacharopoulos, "An Economic Analysis of Labor Skill Requirements in Greece." Ph.D. dissertation, Department of Economics, University of Chicago, 1968.

[8] Joseph A. Kershaw and Roland N. McKean, *Teacher Shortages and Salary Schedules, op. cit.*

(b) Economic cost-benefit studies assume that differences in income that are associated with the number of years schooling completed are the result of education, rather than that both income and education are affected by native ability, motivation, and parental status. It is difficult to deal empirically with this problem because "native ability" is an evasive concept, and is not at all the same thing as I.Q.

(c) For a number of reasons, including (a) and (b) above, the rate of return calculated on the basis of cross-sectional data is not necessarily the rate which would reward additional expenditures in education. Findings such as those reported in Table 6-2 provide clues or hypotheses regarding the probable effect of various levels of resource allocation for education, but they do not guarantee that this effect will in fact come about.

In spite of these limitations the formal models listed above do provide a basis for practical applications to resource allocation problems. In particular, they undergird any theory of budgeting in education which will be developed on economic principles. Before turning to the topic of budgeting, it is appropriate to again examine a nonbudget allocation problem, that of making best use of the time of the personnel within an educational system.

THE ALLOCATION OF TIME

One of their most important contributions to the study of education has been economists' emphasis on the value of students' time. Schultz and others have shown that the foregone earnings associated with students' being in school instead of in the labor force result in large additional costs, which must be taken into consideration in both public and private decisions to invest in education.[9] Becker illustrated the importance of this concept, calling attention to the failure of free tuition to eliminate impediments to college attendance, and to the increased enrollments which sometimes occur in geographic areas which are economically depressed, and in periods of economic depression.[10]

In a recent article, Becker pointed out that there are two ingredients

[9] Theodore W. Schultz, "Education and Economic Growth" in *Social Forces Influencing American Education* (Chicago: University of Chicago Press, 1960)

[10] Gary S. Becker, *Human Capital* (New York: Columbia University Press, 1964) ch. IV.

in the cost of commodities. One ingredient consists of the goods and services which go into the production of the commodity; the other ingredient is the time taken to consume the commodity. This analysis explains, for example, why people may prefer to have milk delivered to their door, even at a higher cost; the value of the time taken to visit the store makes the total cost of delivered milk less than that of milk purchased at the store.[11]

Education is an excellent example of a commodity whose cost includes a large time-related element. Obtaining a college degree entails foregoing earnings and also other activities for a relatively long period of time. If the person's education includes post-graduate training, or the long periods of training and internship associated with the medical earnings, time is indeed a major element of total cost.

While the time-related costs of higher education may consist largely of foregone earnings, there are other types of education whose costs consist largely of foregone leisure. Consider, for example, the cost of attending an evening course designed to enrich the background of students, to improve their work skills, or to provide them with immediate consumption, in the form of an enjoyable way of using their time. Adult educators may despair of their ability to attract students, even when courses are provided at a low rate of tuition. However, course attendance is never free, even if the person is not foregoing gainful employment, or subject to such expenses as transportation or baby-sitting. There is still the foregone opportunity to enjoy leisure time in other ways. For people who place a high value on leisure, the cost of attending evening classes or engaging in civic activities may be prohibitive.

Suppose, however, that a given amount of time has been allocated to education, as in the case of an individual who has enrolled in college. The costs associated with foregone earnings have been taken into consideration and the problem now becomes to make the "best" use of time allocated to education. There are still many choices available, including what courses to take, what extracurricular activities to engage in, and what use to make of uncommitted time. Suppose $Z_1, Z_2, \ldots Z_n$ are the various activities which are considered. The total utility of the mix of activities is designated by U. The problem is to maximize U, which is defined as follows:

$$(1) \quad U = g(Z_1, Z_2, \ldots Z_n)$$

In other words, total utility depends upon the amount of each activity that is "consumed." Thus the Z's (separate activities) directly provide

[11] Gary S. Becker, "A Theory of the Allocation of Time." *Economic Journal*, September, 1965.

utility to the individual. Each Z, however, is also produced by the individual by combining the services of goods or people (such as the services of a teacher, a book, or a television set) with some of his own time:

$$Z_i = f_i(x_i, t_i)$$

It has been shown by economic theorists that the maximizing individual will tend to allocate his time and purchased resources among activities so that the marginal addition to his total utility is the same from each activity. In mathematical language:

$$\frac{\partial U}{\partial Z_1} = \frac{\partial U}{\partial Z_2} = \cdots \frac{\partial U}{\partial Z_n}$$

There are, of course, constraints on both the total goods (i.e., total amount of money available) and, equally important, the total time which is available. Hence, the problem of the administrator and also the individual is to make the best possible use of time, and also of the money at his disposal.

Implications

To summarize, once decisions have been made by individuals or by governmental or private organizations to provide resources for education for a period of say one year, the most important choices of individuals or groups consist in determining how the committed time will be used.

This choice is made as follows. Suppose there are only two alternatives which are called courses A and B. A person has to make up his mind (or an administrator has to schedule students) so as to divide a given amount of time between courses A and B. A number of other variables are involved. For example, students differ in the time it takes them to master the objectives of courses A and B. Individuals or institutions may apply different weights to the importance of mastery of objectives in A and B. Either one of A or B may be more difficult, and hence more time consuming, for all students. The final choice depends on the notion that the cost of spending a unit of time in course B consists of the foregone opportunity to spend this time in course A. Hence, the final decision is based on the notion that time should be allocated so that the value of the marginal hour spent on A (in terms of foregone learning opportunities in B) is equal to the value of the marginal hour spent on B (in terms of the foregone learning opportunities in A).

1. Because there are interpersonal differences in learning rates, and also in the weights placed on the various objectives, decisions should be made by the learner, whenever possible. This argues against requiring certain courses for all college students, and in favor of permitting students to use optional methods (such as private study) to meet a given set of objectives.
2. At the high school level, this analysis suggests that the Carnegie unit system, which requires certain fixed amounts of time to be spent by all students in each course, should be modified. The analysis argues for flexible scheduling procedures, and the allocation of time so as to equalize the marginal product of time spent in various learning activities.
3. Since learning rates differ from student to student, there should be ways to permit students to advance through the grades or to master given bodies of student matter at their own speed.[12]

This discussion may be summarized by reiterating that time is an important commodity for students at all grade levels, the school schedule is an important allocatory mechanism, and the newer procedures that stress flexibility in time allocation provide the opportunity to maximize learning outputs by permitting students, teachers, and administrators to experiment with various production functions.

THE BUDGET AS AN ALLOCATORY TOOL

Careful planning is required if the productivity of educational systems is to be maintained and improved. The planning process includes setting objectives, and identifying alternative procedures for attaining them. It also includes the development of procedures for choosing from among alternative procedures. Since the production of human capital takes place over a relatively long time period, educational planning must comprise a time perspective of a number of years.

Conceptually, the budget is an instrument designed to facilitate planning. It provides a format within which allocatory decisions can be formulated and implemented. The budget imposes a recognition of

---

[12] Benjamin S. Bloom, "Learning for Mastery," *Evaluation Comment,* UCLA Center for the Study of Evaluation of Instructional Programs, I, No. 2 (May, 1968).

the constraints imposed by the limited availability of resources. Within this resource constraint, it requires the identification of specific items of expenditure, and the classification of expenditures into analytic categories.

The budget also provides a context for a planning *process,* or a series of activities in which many types of people can be engaged in selecting the means they will use for reaching a given set of objectives. The budget becomes a *document* which summarizes the planning decisions, in a manner which permits comparisons to be made over time in a given educational system, as well as among systems.

In addition to serving this major purpose, the budget acts as a device for ensuring the careful and honest stewardship of public funds. The budget is a public document that may be studied by people outside the system, thus providing for an external scrutiny of expenditure plans. Formal auditing procedures are subsequently used to compare budgeted expenditures with actual expenditures, and thus to provide maximum safeguards against dishonest or careless practices.[13]

A distinction must be made between the budget as an instrument for ensuring economic rationality, and the actual process by which the budget is constructed. In a real world, inhabited by real people, social, political, and psychological influences are bound to influence the decision-making process. We turn first to a brief discussion of the budget process.[14]

This section presents a paradigm of rational budgeting, and contrasts it with some of the characteristics of budgetary decision making in school systems.[15] This so-called "rational paradigm" is an abstraction, and is not meant to be applied without modification in a given situation. Psychological and interpersonal considerations preclude complete ration-

---

[13] In order that the budget may serve this stewardship role, all funds associated with the school systems should be included in it. This includes funds received from federal as well as state and local sources. The practice which is occasionally used of excluding certain petty cash, athletic or revolving funds from the regular accounting and budgeting procedure is to be deplored, both because of the opportunities for possible misuse of funds which arise, and because such practices exclude part of the total resource allocation process from cost-benefit analysis.

[14] J. Alan Thomas, "Educational Decision-Making and the School Budget," *Administrator's Notebook,* **XII,** No. 4 (December, 1963).

[15] For a similar comparison, see Charles E. Lindblom, "Decision-Making on Taxation and Expenditures," in *Public Finance: Needs, Sources and Utilization* (Princeton: Princeton University Press, 1961), pp. 295–336; and Arthur Smithies, *The Budgetary Process in the United States* (New York: McGraw Hill, 1955).

120 ALLOCATION AND BUDGETING

ality in human organizations.[16] However, the paradigm provides a base line against which departures from rationality can be gauged.[17]

1. In rational budget making, goals or objectives are clearly specified, preferably in operational terms.
2. In the rational model, inputs are selected and combined in such a way as to maximize goal attainment. Scientific procedures such as systems analysis, which involve the study of input-output and cost-benefit relationships, are used as tools in the development of procedures designed to improve the productivity of educational systems.
3. In rational analysis, a number of alternatives are identified and compared before a decision is made to implement a given procedure.
4. In the rational model, information systems are used as a basis for the improvement of decision making.
5. The rational model includes an evaluation of the results of implementing a given budget.
6. The rational model places much emphasis on long-term planning.

Let us compare this model with a procedure typically followed by school systems.

1. Some time in advance of the budgetary deadline, teachers and other school personnel are required to submit lists of supplies and materials needed for the coming year.
2. These lists are consolidated and modifications are made within each school before the lists are submitted to the central office.
3. On the basis of these lists and those submitted by central office personnel, the budgetary requirements of the various suborganizations—instruction, plant maintenance and operation, administration, and special services—are compiled.
4. At a meeting of the administrative cabinet, these subbudgets are reconciled and used in the preparation of a master budget.
5. The master budget as presented to the Board of Education is usually a modification of the previous year's document. Often,

[16] Herbert A. Simon, "A Behavioral Model of Rational Choice," in *Models of Man* (New York: Wiley, 1957), pp. 241–260.

[17] Note: It should be emphasized that behavior characterized by a departure from economic rationality need not be characterized as irrational. For example, actions may be highly rational from the point of view of obtaining political support for a budget, even though the budget itself may not be directed toward the objective of maximizing the defined objectives of an educational system.

where resource increases are asked for, they tend to be more or less equal percentage-wise "across the board," rather than constituting a reallocation among budget categories or programs.
6. The final budget comes about as a result of some hard bargaining among individuals representing various suborganizations.
7. Budgeting tends to be for a year in advance, with little attention to long term planning.

There is a clear lack of correspondence between the so-called rational model and the way in which budgeting is carried out in school systems. The latter procedures, however, have much merit. Typically, many if not most members of the system are involved in the process, and this may improve the degree to which the final decisions are supported and implemented. The bargaining process which is used to determine the manner in which resources are allocated takes into consideration the reality of political power within an educational system. This dependence on incremental decisions in budget preparation is realistic, since large scale resource reallocations are seldom politically possible. Furthermore, an incremental budget is easier to explain to a lay public than a budget developed on the basis of a cost-benefit analysis of the total operation. Some combination of the traditional budgetary procedures with the elements implied in the rational model is clearly desirable. The next section describes the analytic procedure used in current budgetary practice, and elements of the rational model that are emerging in program budgeting procedures.[18]

THE BUDGET DOCUMENT

The educational budget is a statement of planned receipts and expenditures, classified in a manner which improves the degree to which the document can be understood by people within and outside the system. As noted above, it reflects a series of decisions about both revenues and expenditures, as well as about educational programs. In the final document revenues and expenditures must balance, and the educational

[18] For an excellent discussion of budgetary practice in large American school systems see H. Thomas James, James A. Kelly, and Walter I. Garms, *Determinants of Educational Expenditures in Large Cities of the U. S.* (Stanford: School of Education, Stanford University, 1966) pp. 43, 90, 158–194.

program may therefore be constrained by the limited availability of resources.[19]

The usual type of budget document is a financial document reflecting the elements of the educational program. It corresponds to the first production function described in Chapter 2, which regards programs and services as outputs. The sequence of decisions implied is (1) the selection of program offerings; (2) the use of input coefficients to determine the resource requirements, in terms of such inputs as teachers, space, equipment, and materials as well as the total amount of money which is needed to operate the program; (3) the description of revenue sources. If revenues are not adequate to provide for the defined program, the latter must be modified, although some state statutes permit borrowing to finance current operations. In this case, revenues include money obtained through the sale of debentures.

A "quality" dimension is implicitly built into an educational budget. For these purposes, quality tends to be defined in terms of the nature and quantity of the inputs which are utilized. Thus, the input ratios used to determine the number of teachers to be hired will vary from system to system. The more affluent educational systems will operate with lower ratios of students to various inputs (for example, student-teacher ratios of less than 25, large per student space allocation, and so on); the less affluent systems will have higher ratios of students to similar inputs.

Also, if more money is available, salaries tend to be higher. If higher salaries are accompanied by careful recruitment, the systems with more money at their disposal will obtain teachers who are better qualified, and presumably more competent. Such systems will also tend to purchase better equipment and to construct school buildings which may be both more attractive and more conducive to sound education than those constructed in systems with less money at their disposal. All of these factors are built into the budget. When a budget is presented to the public, the astute superintendent and school board will point out these qualitative factors, and will justify increments in expenditures on the basis of qualitative improvements.

Every budget document is based on a system for classifying expenditures. Since the budget has several purposes, it may include several classification systems, each with a unique purpose. The methods of classification should satisfy at least one of the following criteria. (1) A given classification may improve the usefulness of the budget as an instrument

[19] Chris Anthony de Young, *Budgetary Practices in Public School Administration* (Evanston, Ill.: Northwestern University, School of Education, 1932).

permitting control over allocatory decisions for a given period of time in accordance with the policy determined at the institutional level. (2) A given classification may facilitate the rationality of budgetary decisions, by permitting a comparison of the costs and benefits associated with alternative courses of action. (3) A given classification may permit the budgetary document to be more easily understood by laymen and professionals. Four types of classification schemes will be described in this section: (1) budgeting by line item; (2) budgeting by organizational unit; (3) budgeting by functional category; and (4) budgeting by program or performance.

1. Line Item Budgeting

From the systems point of view, the line item budget is a detailed description of inputs to be purchased, and the cost of each. For the purpose of decision making, each item should be coded in a way that relates it to programs or performance objectives. It is then possible to determine in advance the approximate cost of providing a given program.

2. Budgeting by Organizational Unit

Since state, national and local government have different sets of educational objectives, each will have its own budget. The operating unit of government is the local school district; its budget must include funds received from other levels of government. Where these latter funds are categorical in nature (for example, money for improving the education of children from low income families, or for providing vocational education or driver training), state and national objectives must be recognized in the local budget.

Within each school system, also, some budgetary decisions should be decentralized. In school districts which contain many individual schools, consideration must be given to providing each building principal with a degree of fiscal autonomy. Where there are differences from school to school in student background, each school may need to develop its own production function, in order that its productivity and the productivity of the entire system can be maximized. This implies that system-wide allocatory procedures may be inappropriate, and that budgeting should, perhaps, be carried out on a school-by-school basis. The principal may then be held accountable by the school board and his own community for the achievement of objectives which are determined by himself, his teachers, and the parents of his students.

3. Budgeting by Functional Categories

The traditional school budget is based on categories initially proposed by a committee organized by the U. S. Office of Education.[20] The following are the main categories used in this type of educational budget:

Administration
Instruction
Attendance and health services
Pupil transportation services
Operation of plant
Maintenance of plant
Fixed charges
Food services and student body activities
Community services
Capital outlay
Debt service from current funds
Outgoing transfer accounts

These categories are useful for descriptive purposes. However, they do not constitute useful breakdowns for the purpose of decision making, since there is no way of relating the various inputs included in these categories to either programs or performance objectives. In fact, they may impede decision making, by engaging boards and administrators in the partially irrelevant exercise of examining the manner in which resources are allocated among these categories. From the point of view of the school board member or layman who wishes to use the budget to inform him about the manner in which the system is allocating its resources, the above categories may appear to be developed for the purpose of concealing rather than revealing information. Because of these limitations, attempts are under way to introduce a systems-based budget. Such a budget would be based on the Administrator's Production Function, in which the output is courses or services or, in broader terms, programs; or the Psychologist's Production Function, which relates in-

[20] Paul L. Reason and Alpheus L. White, *Financial Accounting for Local and State School Systems* (Washington: U. S. Office of Health, Education, and Welfare, 1957).

puts to outputs. The systems-based budget is called a program budget or, in somewhat different context, a performance budget.

4. Budgeting by Program or Performance

This book does not attempt to describe in detail the nature of program budgeting.[21] Rather, it proceeds from the framework developed in earlier chapters to discuss the conceptual base on which this format rests. To begin with, program budgeting has the following characteristics.

1. The program budget rests on an explicit listing of services to be performed or objectives to be attained.
2. Wherever possible, resources are allocated in order that these objectives may be maximized. For this purpose, scientific procedures such as cost-benefit, input-output, or cost-effectiveness analyses are used.
3. Program budgeting must include procedures for evaluating the degree to which objectives are achieved.
4. Program budgeting includes the allocation of resources and the projection of goals for a period of more than one year. Hence, program budgeting may be thought of as a planning instrument.
5. If properly carried out, program budgeting includes the identification of alternate methods of reaching a given objective or set of objectives. The "best" alternative is selected through some variation of input-output analysis.
6. Program budgeting requires the development and use of sophisticated information systems.
7. The use of program budgeting does not diminish the need for other kinds of budgets such as those in common use today. Computerized information flow systems permit a given piece of information (say, an expenditure or a revenue item) to be entered in more than one budget format.

Although fairly wide agreement would be reached on the above characteristics, there is some controversy as to the exact programs or performance categories which should be included in the program budget. One method of classification is in accordance with the three production functions discussed in Chapter 2.

[21] A thorough and careful treatment is found in Harry J. Hartley, *Educational Planning-Programming-Budgeting* (New York: Prentice Hall, 1963). See also Charles S. Benson, *The Economics of Public Education* (Boston: Houghton Mifflin, Second Edition, 1968).

1. The Administrator's Production Function. If this approach were followed, the budget categories would be specific services, such as instruction in Biology 1. Such specificity is probably not necessary at this point. Rather, we propose budgeting by level, and for certain special subject areas. An example of an output associated with this type of budget would be one student-year of elementary education. This, clearly, is a gross kind of output measure. However, a budget based on these categories does provide the decision maker with the tools to make important allocatory decisions, such as whether a larger portion of the total budget should be spent on preschool education. Furthermore, it provides important information to the public and to the members of the governing board, which is not available in the traditional budget. An example of a budget based on these categories is reported by Charles S. Benson.[22]

2. The Psychologist's Production Function. Although the foregoing analysis introduces additional rationality into the production function, it does not go far enough. Schools are intended to produce behavioral changes, and are not merely service-producing agencies. For some purposes at least, school systems should experiment with budget formats

[22] Charles S. Benson, *The Economics of Public Education,* 2nd Edition (Boston: Houghton Mifflin, 1968), p. 266.

Table 6-3 Program Budget in Performance-Type Format

Target Population. This sub-budget deals with the teaching of reading to 150 kindergarten and first grade children in the attendance area of the present Taft and Roosevelt schools.

Population Characteristics. The average income level in the community is $3500 per family. Seventy-two per cent of the parents are receiving welfare assistance. The present third grade class is reading on the average one grade below normal. This gap increases as students progress through the grades.

Program Objectives. It is desired, over a 3 and 4 year period respectively to raise the performance of these students in reading to grade norms, as indicated in the Stanford Achievement Tests, Reading sub-test.

Available Resources. $30,000 in federal funds is available, in addition to the normal appropriation of $90,000. Specialized consultant help is available from the school system and the nearby university. Additional volunteer help is available in the community.

Alternative Ways of Utilizing These Resources. These alternatives would be identified and analyzed in some detail.

Alternative Selected. This alternative would be carefully costed out, over a 3–4 year period.

Evaluation Procedure. The results of this procedure would be examined, and compared with the results gained by using other treatments. The evaluation results would then be used in the development of more refined program budgets for subsequent years.

which relate resource allocation to the achievement of system objectives. Table 6-3 is an example of such a budget.

3. *The Economist's Production Function.* Finally, there will be some instances where economic cost-benefit analysis can be used as the basis for budgeting a portion of the system's resources. Table 6-4 provides an example of this procedure.

Table 6-4   Program Budget in Cost Benefit Format

Target Population. This sub-budget deals with 100 teen-age high school drop outs, who live in a given city of medium size. These youths are characterized by a dislike for school, a fear of failure, and a resentment of authority. On the other hand, they are alert, intelligent, and in good health.

Program Objectives. It is desired, over an 18 month period, to assist these youths to obtain the basic skills which they will need in order to obtain employment.

Available Resources. About $100,000 is available for this experimental program.

Alternatives Identified. A number of possible ways of providing this training are identified. These alternatives are compared, in terms of their anticipated costs and benefits. Benefits are measured in terms of the expected likelihood of the students completing the course, and obtaining employment, also their anticipated income after employment. The selected alternative is that which maximizes the benefit cost relationship.

Evaluation Procedure. An ex-post evaluation of the procedure is conducted, including a study of post graduation employment and income. This information is used for studying new proposals for investment in training for employability.

These examples suggest that program budgeting must rely heavily on operations research activities as described in this book. The availability of information about input-output and cost benefit relationships is essential, if available alternatives are to be identified and compared. Operations research should include careful evaluation of applications involving a wide variety of alternative approaches to a given problem. Both cross-sectional and longitudinal studies are essential, if improved practice is to result from such studies.

Program budgeting requires large quantities of information. Computerized information systems are now being developed in both large city school systems and state departments of education. Information should be gathered and stored within a conceptual framework which permits studies to be made of the relationships between inputs and a variety of outputs. Such information systems will cut across traditional lines of demarcation among organizational subsystems.

AUTHOR INDEX

Levin, Henry M., 20, 39, 86
Lindblom, Charles E., 119
Luce, Duncan R., 73

McKean, R. N., 5, 16, 114
Main, Earl D., 88
March, James G., 58, 62
Marshall, Alfred, 5
Meyer, Donald L., 73
Moses, Lincoln E., 73, 95
Mushkin, Selma J., 36

Pederson, K. George, 6
Psacharopoulos, George, 114

Raiffa, Howard, 73, 95
Ramo, Simon, 103
Reason, Paul L., 124

Rogers, David, 104

Samuelson, Paul A., 108
Schultz, Gheordore W., 4, 32, 63, 75, 110, 114, 115
Simon, Herbert A., 62, 100, 120
Smith, Marshall S., 20
Smithies, Arthur, 119
Stigler, George J., 104
Strevell, Wallace H., 112

Thomas, J. Alan, 21, 37, 106, 119
Todd, Edward S., 92

Watts, Harold W., 20
White, Alpheus L., 124
Wilkinson, Bruce, 6
Wolf, Richard M., 14

Subject Index

Administrator's Production Function (PF1), 12
Aspirations (organizational), 62

Benefit-Cost Analysis, *see* Cost-Benefit Analysis
Budget and budgeting, 108, 118–127

Costs, 31–55
   direct, 33, 37
   indirect, 33, 38, 75
   interest, 33
   monetary, 35
   non-monetary, 35
   private, 34
   social, 34
   unit, 42–45
Cost analysis, 31–55
Cost-benefit analysis and allocation, 82(fn.), 87, 113
Cost-effectiveness analysis, 82

Depreciation, 36
Dimishing marginal returns, 67
Direct costs, *see* Costs
Diversification, 51

Economics of education, 1–6

Economics of scale, 45–50
Economists's Production Function (PF3), 22
Environment, effects on learning, 17 ff
Expectations (organizational), 62–63

Feedback models, 98

Indirect costs, *see* Costs
Indivisibilities, 52
Information systems, 103–107
Inputs, 10
Input-output analysis and allocation, 10, 80, 111–115
Interest costs, *see* Costs
Iso-product curve, 70–72

Joint products, 11

Linear programming, 89

Marginal analysis, 63
Monetary costs, *see* Costs

Non-monetary costs, *see* Costs

Objectives, selection of, 75–78
Open systems, 10–30
Opportunity costs, 31

131

SUBJECT INDEX

Output, 80; *see also* Input-output analysis; Objectives, selection of
Output mix, 50–55

Present value analysis, 22
Private costs, *see* Costs
Probability models, 95–98
Production function, 10 ff
 administrators' (PF1), 12
 economists' (PF3), 22
 psychologists' (PF2), 13
Production surface, 67
Program budget, 125–127
Psychologists' Production Function (PF2), 13

Rate of return analysis, 24
Reservation rate of return, 24

Social costs, *see* Costs
Specialization, 53
Subjective probabilities, 73
System, 9

Teachers' salaries, 5–6, 25–30
Time, allocation of, 31–32, 40–41, 63–65, 67–69, 75, 115–117

Unit costs, *see* Costs

Value added, 13, 57
Vocational education, 92–93